Chris Lukhaup and Reinhard Pekny · Invertebrates

Chris Lukhaup/Reinhard Pekny

# INVERTEBRATES

Shrimp, crayfish, crabs & snails in freshwater aquaria

Dähne Verlag

**Photo credits**
All photos courtesy of Chris Lukhaup, if not mentioned otherwise

ISBN 978-3-944821-06-1


Editorial Design: René Türckheim
Translation: Ulrike Bauer
Corrections: Kerry Williamson
Printed in Germany

# CONTENT

**Chris Lukhaup** **Reinhard Pekny**

# FOREWORD

Twenty years ago, our enthusiasm for freshwater invertebrates was triggered by crayfish. In the following years, we went on numerous expeditions searching for them, and we discovered many other invertebrates along the way. Shrimp, anomurans, crabs, snails and mussels soon cast their spell over us as well.

Many aquarium keepers became interested in these so far rather unknown inverts, and pet stores soon started offering a constantly growing, and more varied range of these fascinating animals. In the wake of this trend, people kept asking for closer identification of species, requiring more information about the conditions they needed to create for keeping and breeding these inverts in their aquaria. In our aquaria and breeding tanks we have kept hundreds of species ourselves, and we have bred many of them. Not all the inverts we collected or acquired through the trade turned out to be suitable for a decorative aquarium. Some species require more or less specialist care, but others are great for hobbyist keepers, too.

Unfortunately, we are unable to present in this book each and every species that has fascinated us in the course of time. However, we are proud to present to you the most beautiful and relevant ones. One of our main criteria is that the species can be kept in a tank with relative ease. It is our hope that this book will gain lots of new fans for this fantastic hobby as they succumb to the fascination of freshwater inverts – just the way we did.

**Chris Lukhaup and Reinhard Pekny**

# INTRODUCTION

Back in the eighties, invertebrates were but exotic guests in freshwater tanks, with the only exception being a few snail species, which were usually introduced in aquaria as uninvited guests, often on plants, and one or the other shrimp, almost always called glass shrimp, that was used as live food for fish. For a few years now, though, inverts have been booming. Today we have a varied range of crayfish, snails and – most of all – shrimp, and almost every week a new variant or species is brought onto the market. Some of them establish themselves in the hobby, whilst others disappear from the wholesalers' lists as fast as they entered them. More often than not, unknown requirements for keeping and breeding them coupled with high loss rates and a lack of offspring are responsible for this phenomenon. Moreover, some invert species turn out to be quite incompatible with other aquarium inhabitants, and thus soon cease to grace the aquaria at the dealer's place.

Their great variability makes it hard to describe all known species in detail, however, with this book we are trying to give our readers an overview of the most popular and the most common species available, and we hope to be able to help prevent mistakes in keeping and breeding them, and the disappointment that follows mishaps. We have not described every colour morph in detail, as the requirements of variants of one species are usually more or less identical. However, we pride ourselves in showing the vast variability of nature and deliberate breeding in captivity by including a large number of additional photos.

The basic precondition for keeping any animal in captivity is the exact knowledge of the species and its requirements. Unfortunately, gross mistakes keep happening, some of them as early as at the importer's or wholesaler's, and wrong information is passed down to the customer. Admittedly, it is not always easy even for specialists to define a species reliably without performing an exact, elaborate scientific determination. Some crustacean species are often mistaken for each other in trade. Sometimes, shrimp are even sold as crayfish – and vice versa –, and the outlook is not much brighter when it comes to crabs or anomurans.

The information provided in the following will enable you to differ clearly between the groups of decapod crustaceans living in fresh water.

All decapod crustaceans have five pairs of walking legs, with which they can support themselves and more. The form of these legs enables us to assign these inverts to their individual group, easily and unmistakably.

### Shrimp

The tips of the first two walking leg pairs carry pincers, or chelae. There are species in which both pincer pairs have been transformed into fans. If there are no fans, the chelae on the second leg pair are always the stronger ones.

### Crayfish

There are chelae on the tips of the first three walking leg pairs. The chelae on the first pairs always form large "crayfish claws". The second and third leg pairs carry simple small pincers, the remaining two walking leg pairs end in a thorn.

### Anomurans

The first and fifth walking leg pairs end in claws. The first pair has large chelae, the fifth is always considerably smaller and ends in tiny pincers. As a rule, you can usually only see the large chelipeds, followed by three more walking leg pairs. The smaller leg pair with pincers is hidden away under the pleon or tail, which is folded in half under the carapace.

### Crabs

Here only the first walking leg pair is equipped with chelae, the remaining legs end in thorns. The pleon is folded entirely under the carapace.

## Snails

Like crustaceans, freshwater snails (mollusks) have gained the increased attention of hobby aquarium keepers during recent years. Nowadays you will find numerous and very attractive species in trade.

## Mussels

Another mollusk group consisting of the bivalves, or mussels.

# SHRIMP

SHRIMP

56 SPECIES

# SHRIMP

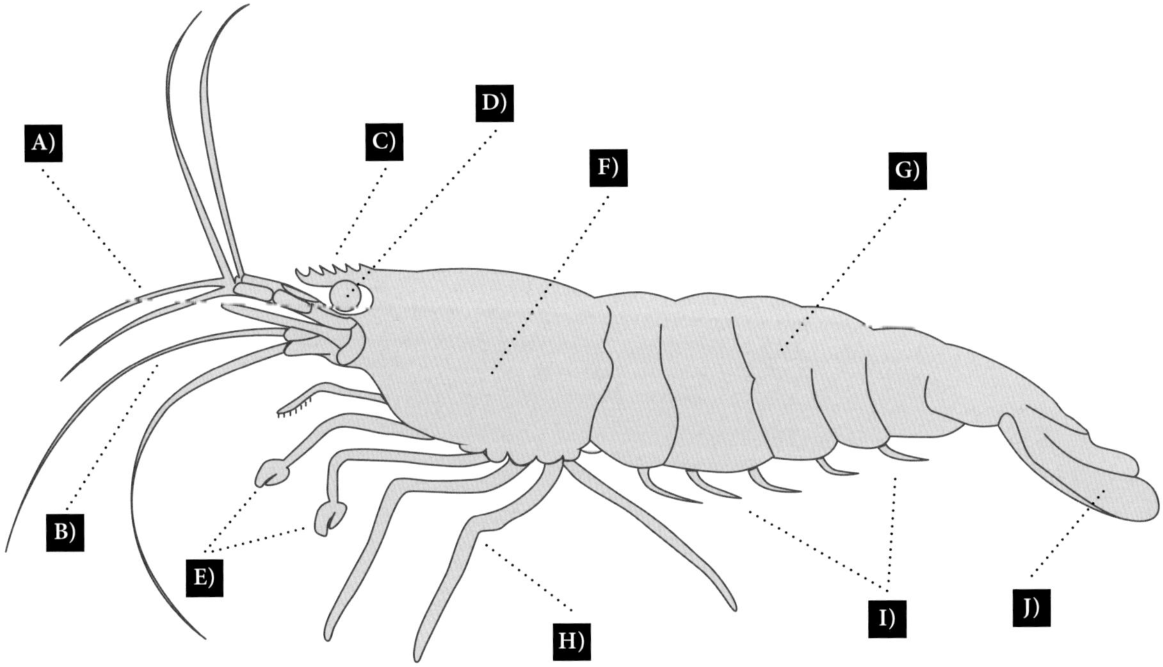


# ANATOMY

A) first antennae pair
B) second antennae pair
C) rostrum
D) eye
E) claw
F) carapace (front)
G) abdomen/pleon
H) walking leg
I) swimmerets/pleopods
J) uropods (tail fan)

Since the late 1990s, ornamental freshwater shrimp have been an integral part of the hobby. Before that turning point, pet stores sold them from time to time, usually as live food. But then, a veritable boom set in, and today shrimp may even be the most popular inverts in our aquaria.

The shrimp common in our hobby belong to different genera and families, scientifically speaking, but what unites them is that they spend all of or at least the most part of their lives, especially as adults, in fresh water.

Some species have not become entirely independent of the original habitat of their ancestors, the sea, and they need brackish or marine water for reproduction. These species belong to the so-called primitive type and produce large numbers of very small eggs per batch. The larvae hatching from these eggs are released into the open water, where they form part of the plankton and go through many stages of development. Only towards the end of their time as larvae do they start a benthic life on the ground. Around this time, they also migrate back to pure fresh water.

During the course of evolution, freshwater dwarf shrimp have spread from the estuaries of larger rivers to rapidly-flowing creeks in the mountains, to enormous lakes, the smallest trickles of water and even to cave waters. They are found from tropic to temperate climates today, however, they have not conquered the arctic regions yet. The abundance of different habitats has resulted in a great variability in shrimp species and in stunning forms. Their sometimes truly impressive colours and patterns are the result of their adaptation to the different living conditions in their habitats.

From all the different shrimp in the world, only representatives from three groups have managed to find their way into our aquaria - dwarf ornamental shrimp, fan shrimp and long-arm shrimp. They differ in body size and form as well as in their habits. The requirements regarding their environment do not differ much between shrimp belonging to one of these groups. Practically every shrimp available in the trade belongs in one of the three groups, under systematical aspects. Dwarf shrimp are probably the most prominent and also the most popular among them. They have found their way into the aquaria and into the hearts of their keepers in allmost all the world.

The shrimp offered in the trade today are rather variable in size. Dwarf shrimp with a total body length of around 15 mm and long-arm shrimp of the genus *Macrobrachium*, some of whose representatives can grow to a body size of over 50 cm in length, of course differ considerably when it comes to their keeping requirements. For this reason it is practically impossible to describe a "typical" shrimp tank – you'll have to decide which species or at least group of species you want to keep. Dwarf ornamental shrimp can be kept in tanks as small as 20 litres upwards. Smaller tanks are available in the trade, however, they bring about some grave disadvantages. The smaller water body reacts too fast to temperature changes in its surroundings, and it is difficult to achieve stable water parameters in there.

## Filtration

Depending on the type of shrimp, their requirements regarding filtration differ considerably, too. Tanks dedicated to dwarf shrimp can be equipped with small internal filters, air-driven sponge filters or undergravel filters, for example. A slight flow is preferable. Even though many dwarf shrimp species originate from small creeks it is not necessary to provide them with a strong current. On the contrary, a strong filter always carries the risk of shrimp being sucked into its intake. A strong current in a shrimp tank is only possible if the filter intake surface is sufficiently large, like that of an undergravel filter or a large-dimensioned filter mat, for example. This kind of filtration and filter intake is especially suitable for shrimp tanks with fan shrimp. Most fan shrimp originate from rapid or even torrential creeks, and they just love sitting in a good, strong current where they can catch food from the water column with their filtering fans. The risk of them being sucked into a filter intake is practically nil, however, their food will be removed from the water too fast if the filter intake current is overly strong. The keeping requirements of long-arm shrimp can be compared to those of crayfish. A well-dimensioned strong filtering system, be it internal or external, is the way to go.

## Decoration

Most shrimp that can be kept in an aquarium are good swimmers. Unlike fish, however, they don't need open water to swim but prefer a really well-structured tank providing them with various surfaces to sit on. For this purpose you can use all kinds of aquarium decoration. Es-

pecially dwarf shrimp will even use not overly clean vertical glass panes of your aquarium to sit on, given that a little algae and biofilm have had the chance to form on them. Fan shrimp need a sufficient number of places to sit, depending on the size of their group, where they can make full use of the current and still find some shelter. If there is an insufficient number of such places in the tank they will even start fighting for a good place. Long-arm shrimp need a well-structured aquarium, especially if you are planning on keeping more than one (not possible for all long-arm shrimp species) or breeding them. As these larger shrimp are not really great swimmers, especially when they grow bigger, it is important to provide them with three-dimensional structures in the tank where they can climb. They also like caves and hideaways under wood and rocks to which they can retreat.

## Lighting

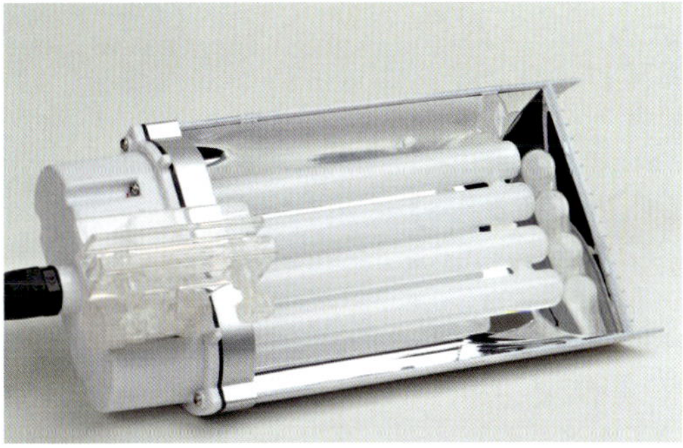

In a shrimp tank, light does not only influence the behaviour of some shrimp species but also the formation of algae and microorganisms. These are important parts of the everyday diet of most dwarf ornamental shrimp, and thus your lighting system ought to be well-adapted to the species you want to keep. If the behaviour of your shrimp tells you they find their tank too bright you can use floating plants to somewhat diffuse the light in the tank without having to invest into a new lighting system.

## Plants

Most shrimp do not eat aquatic plants, thus you can plant your tank just as you please. Even though many shrimp originate from waterbodies with no higher plant growth, they do not mind living in a densely planted tank at all. In a tank dedicated to fan shrimp please make sure these

somewhat plumper shrimp still have room to move without hindrance, though. They clearly prefer unplanted areas with rocks or stones.

## Growth and moulting

Shrimp can only grow when they shed their old, inflexible skin. The physiological processes are the same in all crustaceans, however, the moult

itself goes faster the smaller the shrimp is. Some shrimp even shed their skin while swimming, and you hardly even notice that they moulted just now. The larger a shrimp grows the more exigent the process becomes. In large specimens of the genus *Macrobrachium* it can take as long as it takes a river crayfish of comparable size.

## Diet and Food

Most dwarf shrimp will gladly eat almost all sorts of food you can buy in the aquarium trade. Meanwhile many products have been brought onto the market that have been specifically tailored to the needs of ornamental shrimp. Besides artificial food, a large part of a shrimp's diet consists of things that grow in the tank itself, as they graze intensively on all kinds of surfaces, eating algae and other microorganisms. Failure to re-

move a few dead plant parts or putting dry autumn leaves from the forest (e. g. from oaks, willow trees or alders) into your shrimp tank will be quite beneficial. The detritus forming from them is the main staple food of shrimp in their natural habitats.

Fan shrimp are filter feeders and thus need special food and a different type of set-up, which is described more closely in the species descriptions.

Female

Male

# DWARF SHRIMP

With over 290 species, shrimp of the genus *Caridina* are one of the most diverse groups within the *Atyidae* family. They are also the most widely spread shrimp in the aquarium hobby. However, recent research has found that this genus is in urgent need of a scientific review and re-structuring as there are many discrepancies to be found. The genus *Neocaridina* has up until now been represented by 30 species, and has also found wide distribution in the hobby.

## *Caridina logemanni* "Crystal Red"

Crystal Red Shrimp, Red Bee Shrimp

It is the undisputed queen of all shrimp, and with its myriad of colour morphs and patterns it has become the most popular freshwater shrimp in the aquarium hobby ever. The red colour morph is said to have been discovered by a Japanese shrimp enthusiast, Hisayasu Suzuki, in one of his shrimp tanks in 1991. By selective breeding and back-crossing he managed to get a true-breeding strain, and thus he laid the basis for their victory march around the world. Being the good businessman that he was, Mr Suzuki had the name, Crystal Red, protected in Japan, leading to other names such as Red Bee or Red Zebra also being used for these shrimp. The final breakthrough for the Crystal Red Shrimp came in 2001, when Minoru Yamagishi showed his bright red-and-white shrimp in a photo destined to enter the planted tank contest organised by ADA. This led to a global run for these shrimp, everyone wanted to have them. Even the high prices that were originally paid for them did not halt their rapid spread in the aquarium hobby. As the red-and-white colour variant is a captive-bred form we need to have a look at their wild counterparts, the Bee Shrimp, to learn something about the conditions they prefer. Depending on the degree of selective breeding these shrimp can be more susceptible than their wild predecessors. Some strains have been bred in extremely soft water for years and will react strongly to different water parameters when put into another tank.

Body size: 20 to 30 mm (0.8 to 1.2")
Temperature: 16 to 24 °C (61 to 75 °F)
pH value: 6 to 7.5
Hardness: up to a KH of 10
Tank size: 20 litres (5.2 US gallons) and more
Origins: Tank-bred colour variant from Japan
Reproduction: in fresh water

# CLASSIFICATION

## K CLASSES AND INTERNATIONAL BEE SHRIMP GRADES

### Original Form K0

Grade C

The original form of the Crystal Red Shrimp, with very small white parts. This applies to the red as well as to the black-and-white wild form.

### Red Bee K2

Grade B

Grade B or K2 is the name for the original form with larger white parts.

### Red Bee K4

Grade A

This variant has three or four white bands. It is irrelevant whether the tail has some white, too.

### Red Bee K6

Hinomaru/No Entry Sign

Hinomaru is the name for shrimp whose central coloured band is reduced to a mere circle, and a No Entry Sign is characterised by a white horizontal bar across that circle.

### Red Bee K10

Grade SS/Double Hinomaru, No Entry Sign and Half Moon

Shrimp with this pattern have reduced red bands that do not go all the way round any more but form circles or circular patterns.

### Red Bee K12

Grade SSS/Mosura/Mosura Crown/Mosura Heart

These Red Bee Shrimp have almost no or only minimal black or red colour left on their pleon. Ideally, the pleon is completely white.

### Red Bee K14

Mosura Flowerhead, Mosura Lightning, Mosura Smiley

The patterns on the heads of these Mosura resemble flowers (Flowerhead), flashes of lightning (Lightning) or a smiley face (Smiley). There are more patterns, too, all of which are still called K14.

### Snow White/Golden Bee/White Shrimp

The Snow White Shrimp is all white. For a long time, it was considered to be the perfect Bee Shrimp. Nowadays, there are different classifications, too.

### Super Crystal Red

This is the name for shrimp that are mostly red, with almost no or only a few white parts.

### Pure Red Line

This is a special Red Bee Shrimp line that has remained free from Snow White Shrimp influences. Some breeders crossed Snow White Shrimp with their Red Bee lines in order to improve the proportion and the quality of the white parts of their shrimp. Other breeders, however, think that, quality-wise, the exact contrary was the outcome and that it is better to improve the quality and colour density of the white parts by careful selective breeding. The shrimp resulting from this approach are called Pure Red Line. At the moment, PRL shrimp are the highest-priced shrimp in the market, with prices up to 1,000 $ and more per shrimp.

**In additon to Red Bee Shrimp, the names of the patterns of course also apply to the black-and-white Black Bee Shrimp.**

K0

K2

K4

K6

K8

K10

K12

K14

Snow White

Super Crystal Red

No Entry

Maro-Ten

K0

K2

K4

K6

K8

K10

K12

K14

K14 BB

Spotted Head

No Entry

Maro-Ten

SHRIMP

## *Caridina logemanni* "Bee"

**Bee Shrimp**

With its black-and-white pattern and the orange to yellowish tail fan, the Bee Shrimp can easily be discerned from other shrimp species. It is quite variable in colour distribution and pattern. There are some with densely coloured black and white parts as well as those with a washed-out yellowish white and different shades of grey, and everything in between. Colour intensity depends very much on the well-being of the shrimp and the strength of the lighting on the tank. The Bee Shrimp is the

wild ancestor of the tank-bred Crystal Red Shrimp and originates from only three creeks in the vicinity of Hong Kong. The Bee Shrimp used to be called *Caridina* cf. *cantonensis*, but has recently been described as *Caridina logemanni*. The habitat where Bee Shrimp are actually found has a dense vegetation on the banks of the creeks, and the waters are relatively cool, with a rapid current. The creek bottom consists of rock, with accumulations of dead leaves. In March, during rainfall, we measured a water temperature of only 16.6 °C (61.9 °F). However, the water bodies are subject to considerable changes in temperature in the course of the year, and during the summer months the water may reach temperatures of up to 24 °C (75 °F). In the aquarium, Bee Shrimp can be kept without a heater. If temperatures drop below 18 °C (64.4 °F), they will stop reproducing. The Bee Shrimp lives exclusively in fresh water, and the females produce only a few but rather large eggs.

Body size: 20 to 30 mm (0.8 to 1.2")
Temperature: 16 to 24 °C (61 to 75 °F)
pH value: 6 to 7.5
Hardness: up to a KH of 10
Tank size: 20 litres (5.2 US gallons) and more
Origins: Hong Kong
Reproduction: in fresh water

## *Caridina mariae* "Tiger"

**Tiger Shrimp**

Several different shrimp in the trade are called "Tiger Shrimp". Tiger Shrimp have recently been described as *Caridina mariae*. Only the Camouflage Tiger Shrimp belongs to the Bumblebee Shrimp (*Paracaridina meridionalis*). Tiger and Bee Shrimp interbreed but do not belong to the same species. Both belong to the species group around *Caridina serrata*. The wild forms of the Tiger Shrimp have characteristic vertical stripes on their pleon, which remind of a tiger pattern.

Depending on the location where the animal was originally collected, these stripes may be thicker or thinner. The colour of the tail fan and the head carapace may also be different. In the aquarium hobby, though, several colour morphs have been established, among them the uniformly Black Tiger Shrimp, Blue and Red Tigers. The wild forms all come from southern China, where they are collected in creeks and on flooded grassland. If you mimic the natural temperature curve when keeping them in an aquarium they can be highly productive and will have considerably more offspring than Bee Shrimp. Room temperature is sufficient, however, keep in mind that Tiger Shrimp do not like too high temperatures during the summer months.

Body size: 20 to 30 mm (0.8 to 1.2")
Temperature: 15 to 26 °C (59 to 79 °F)
pH value: 6 to 7.5
Hardness: up to a KH of 10
Tank size: 20 litres (5.2 US gallons) and more
Origins: southern China
Reproduction: in fresh water

SHRIMP

## *Caridina mariae* "Black Tiger"
Black Tiger Shrimp

There are several classes of Black Tiger Shrimp, from transparent with a black pattern to a uniform intense black. In entirely black shrimp, the tiger pattern disappears totally. Some specimens have light-coloured, orange eyes, the eyes of others are dark. Here you also have grades and classifications, depending on the percentage of the coloured parts and the eye colour. Uniformly black shrimp without any transparent spots and with black antennae and black legs are considered the most valuable ones. This colour morph was first bred in Germany, but has spread to almost all the world by now.

Body size: 20 to 30 mm (0.8 to 1.2")
Temperature: 15 to 26 °C (59 to 79 °F)
pH value: 6 to 7.5
Hardness: up to a KH of 10
Tank size: 20 litres (5.2 US gallons) and more
Origins: tank-bred colour variant from Germany
Reproduction: in fresh water

SHRIMP

## *Caridina mariae* "Blue Tiger"

Blue Tiger Shrimp

Among the Blue Tiger Shrimp, you can find specimens with light or dark shades of blue, with light- or dark-coloured eyes. Those with light-coloured eyes are considered a different breed, they are called Blue Tiger OE. OE stands for "orange eye". Under stress, Blue Tiger Shrimp may lose their blue shade and become totally transparent. The Blue Tiger colour morph seems to be somewhat more productive than the other colour variants of the Tiger Shrimp, however, they are also a bit more susceptible to negative influences in their environment.

Body size: 20 to 30 mm (0.8 to 1.2")
Temperature: 15 to 26 °C (59 to 79 °F)
pH value: 6 to 7.5
Hardness: up to a KH of 10
Tank size: 20 litres (5.2 US gallons) and more
Origins: tank-bred colour variant from Germany
Reproduction: in fresh water

## *Caridina mariae* "Red Tiger"

Red Tiger Shrimp

The beautiful Red Tiger Shrimp are much less common. From time to time, red-striped specimens can be found among shrimp collected from the wild in southern China. In the meantime, some breeders have specialised in Red Tiger Shrimp, and we have different classifications here as well, all of them depending on colour density. There are strains that have been bred with Red Bee Shrimp. These Red Tigers should be called *Caridina mariae x logemanni*.

Body size: 20 to 30 mm (0.8 to 1.2")
Temperature: 15 to 26 °C (59 to 79 °F)
pH value: 6 to 7.5
Hardness: up to a KH of 10
Tank size: 20 litres (5.2 US gallons) and more
Origins: China
Reproduction: in fresh water

## *Caridina mariae* "Super Tiger"

### Super Tiger Shrimp

The Super Tiger Shrimp is a rare wild form with broader black stripes on the pleon and a yellow to orange head and tail fan. In nature, these shrimp are not found in the cool waters of a creek but on swampy grassland near the city of Conghua in southern China. They belong to the species group around *Caridina serrata*. Meanwhile, they have been described as *Caridina mariae*. In some places, they occur together with *Caridina cantonensis* without crossbreeding. In the aquarium they can be kept at room temperature. The shrimp do badly at higher temperatures, above 26 °C (79 °F). Like most dwarf shrimp, the Super Tiger Shrimp prefers water with a very low conductivity.

Body size: 20 to 30 mm (0.8 to 1.2")
Temperature: 15 to 26 °C (59 to 79 °F)
pH value: 6 to 7.5
Hardness: up to a KH of 10
Tank size: 20 litres (5.2 US gallons) and more
Origins: southern China
Reproduction: in fresh water

# Crossbreeds

**Red Tiger hybrid**
**(Crystal Red x Tiger)**

Tigerbees are the hybridised offspring of a Tiger Shrimp and a Bee Shrimp. New patterns and colour variants like this are especially interesting for breeders. However, these patterns still need more selective breeding and stabilisation.

**Black Tiger hybrid**
**(Bee x Tiger)**

A crossbreed between a black-and-white Bee Shrimp and a Tiger Shrimp results in a very special pattern.

**Red Tiger Crystal**
**(Red Tiger x Crystal Red)**

Crossbreeds between Red Tiger Shrimp and Crystal Red Shrimp have a denser red colour and wider red stripes. There are almost no white pigments any more.

**Bumblebee Bee**
**(Bumblebee x Bee)**

This cross has been the reason for heated discussions among shrimp breeders for a long time. For a long time, it was disputed whether these two species could be crossbred at all. In Germany it was proved that it is indeed possible, down to the F2 generation.

**Fancy Tiger**

This breed also comes from Germany, and according to the breeder, it is the result of crossing, selective breeding and backcrossing between the following colour variants: Crystal Red, Snow White, Red Tiger Shrimp (collected from the wild), Bee Shrimp, Super Red Tiger Shrimp and the mixes thereof. All in all, this process has taken over three years. Fancy Tiger Shrimp come in red and white as well as in black and white.

## *Caridina* sp.

Tupfel shrimp

Under the name "Tupfel Shrimp", several different species have been imported, however, it is still not quite clear exactly which species they belong to. What is clear is that these shrimp also come from the south of China and that they are very variable in colour and pattern. In the aquarium, they ought to be provided with the same conditions as most other dwarf shrimp from the cool soft waters of China.

Body size: 20 to 30 mm (0.8 to 1.2")
Temperature: 15 to 26 °C (59 to 79 °F)
pH value: 6 to 7.5
Hardness: up to a KH of 10
Tank size: 20 litres (5.2 US gallons) and more
Origins: China
Reproduction: in fresh water

SHRIMP

## *Caridina* cf. *babaulti* "Green"
Green Shrimp

Their name is quite misleading, as these shrimp may be transparent, light brown, greenish-blue, orange or yellow in colour. Differently-coloured shrimp are, as a rule, females that are either berried, ready for mating or more advanced in age. This colour is by no means stable and may change from one minute to the other. Sometimes the Green Shrimp is thus also called Rainbow Shrimp. Some specimens have a striking lighter-coloured stripe along their back, others don't.

These shrimp need a long time until they adapt to a tank, and they are more difficult to breed than other dwarf shrimp. For breeding, and especially for raising the offspring, the tank needs to be very well-planted. Do not siphon off the substrate too often, as the young shrimp of this species need lots of detritus and the microorganisms detritus contains. The Green Shrimp is rather shy and lives a relatively secluded life inside the aquarium, however, they are highly suitable for social tanks holding other dwarf shrimp. Recent research has shown that the Green Shrimp is really a scientifically undescribed species. They are currently under description.

Body size: 15 to 30 mm (0.6 to 1.2")
Temperature: 20 to 31 °C (68 to 88 °F)
pH value: 6.8 to 8.3
Hardness: up to a KH of 25
Tank size: 20 litres (5.2 US gallons) and more
Origins: India
Reproduction: in fresh water

## *Caridina* cf. *babaulti* var. "Stripes"

Indian Zebra Shrimp

Among the normal *Caridina* cf. *babaulti* "Green" imported by wholesalers, from time to time you can find shrimp with a much darker colour and a black striped pattern on the pleon. These shrimp probably also come from India, and they are most likely collected in the same habitat as the green *Caridina* cf. *babaulti*. They probably belong to a different species, though. Their body is much more compact, their rostrum has a different form, and they tend to stay somewhat smaller. According to some specialists, these shrimp are morphologically identical with *Caridina babaulti* sensu Bouvier.

Body size: 15 to 25 mm (0.6 to 1")
Temperature: 20 to 31 °C (68 to 88 °F)
pH value: 6.8 to 8.3
Hardness: up to a KH of 25
Tank size: 20 litres (5.2 US gallons) and more
Origins: India
Reproduction: in fresh water

## *Caridina* cf. *babaulti* "Malaya"

**Malaya Shrimp**

These shrimp are very variable in colour and can be red, blue, purple or even black. Some have a longitudinal stripe along their back, others don't. Many have white vertical stripes across their back and on their pleon. In the trade, these small shrimp are sold

under the name of *Caridina* cf. *babaulti* "Malaya", however, their different body size, their differently formed rostrum and their colours, which are not very typical of *Caridina* cf. *babaulti*, have led us to believe that these shrimp may also belong to a new species in its own right.

Body size: 15 to 20 mm (0.6 to 0.8")
Temperature: 20 to 31 °C (68 to 88 °F)
pH value: 6.8 to 8.3
Hardness: up to a KH of 25
Tank size: 20 litres (5.2 US gallons) and more
Origins: India
Reproduction: in fresh water

## *Paracaridina meridionalis*

**Camouflage Tiger Shrimp, Mustang Shrimp, Bumblebee Shrimp, Larry Shrimp**

Approximately five to six different species are sold under the name "Bumblebee Shrimp" in the trade. The first Bumblebee Shrimp species sold in Germany and Europe in general was, at first, wrongly called *Caridina breviata*. In nature, it is found in oxygen-rich creeks that stay quite cool even through the hot summer months. Besides shrimp with black and white bands, unicoloured specimens, shrimp with stripes or totally different atypical patterns are found. Among the intensively coloured shrimp, you also find many with weak and almost transparent colours, and everything in between. Older shrimp are more susceptible to stress and thus often show weaker colours. In the aquarium, these shrimp reproduce readily. They do not need a heater. During the summer care must be taken as they react badly to high water temperatures, though. The species' high variability and its many different colours and patterns have resulted in several trivial names for *Paracaridina meridionalis*. One variant is known as Bumblebee Shrimp, as it has the typical banded pattern, another striped variant almost looks like a Tiger Shrimp. Often you can find shrimp with a dotted pattern, or cream-coloured ones with a blue basic colour. Some individual specimens may be blue all over, or even black. Very probably, the shrimp are collected in different spots in nature but go into one single tank at the wholesaler's in China. In the aquarium, all variants can be kept in the same conditions as other Bumblebee Shrimp as long as you make sure the water does not get too warm. In their natural habitat, these shrimp mainly feed on dead tree leaves accumulating in heaps in pools with almost no current.

Body size: 18 to 25 mm (0.7 to 1")
Temperature: 17 to 24 °C (63 to 75 °F)
pH value: 6 to 6.8
Hardness: up to a KH of 3
Tank size: 20 litres (5.2 US gallons) and more
Origins: southern China
Reproduction: in fresh water

## *Caridina venusta*

### New Bee Shrimp

Far more rarely found in the trade is *Caridina venusta*, which also inhabits narrow soft water creeks only reaching about 18 to 22 °C (64.4 to 72 °F) during the day even through the hot summer months. Visually, these shrimp are hard to discern from other Bumblebee Shrimp species, as they vary quite a bit in colour intensity and in pattern. According to leading scientists, the examination material might have been mixed up with that of *Caridina maculata* during the first description. The species was previously named *Caridina tumida*, but has been synonimised with *Caridina venusta* as it was proven that both belong to the same species. When *Caridina tumida* was first described, there was some confusion, as some data and characteristics may also have been mixed up. Now these mix-ups have been resolved – all of these shrimp belong to the species *Caridina venusta*. In the aquarium, they need the same parameters as other Bumblebee Shrimp. They love cool, soft water rich in oxygen. In the creeks in their habitat, there are heaps of dead fallen leaves, which are the main food source for these shrimp. Usually, in the shallow trickles of water and the small creeks, there are no fish.

Body size: 18 to 25 mm (0.7 to 1")
Temperature: 17 to 24 °C (63 to 75 °F)
pH value: 6 to 6.8
Hardness: up to a KH of 3
Tank size: 20 litres (5.2 US gallons) and more
Origins: southern China
Reproduction: in fresh water

## *Caridina gracilirostris*

### Pinocchio Shrimp, Long Nose Shrimp, Red Nose Shrimp

Its typical red rostrum and the red basis of the swimmerets make the Red Nose Shrimp easy to discern from any other species. Even though it may be sold as a freshwater shrimp by the aquarium

trade, it is almost exclusively found in brackish water zones of larger rivers and in coastal swamps in nature. It is thus a typical brackish shrimp, which is only rarely found in pure fresh or marine water. As *Caridina gracilirostris* is an avid swimmer, a suitable tank needs a lot of free water zones and thus should not be planted too densely. These shrimp have not been bred in pure fresh water yet, and even just keeping them in fresh water causes problems after as little as six months. Keeping and breeding is quite easy in brackish water, though.

Body size: 30 to 40 mm (1.2 to 1.6")
Temperature: 22 to 29 °C (72 to 84 °F)
pH value: 6 to 8.5
Hardness: up to a KH of 25
Tank size: 40 litres (10 US gallons) and more
Origins: Japan, Indonesia, West Africa
Reproduction: larvae need brackish to marine water

## *Caridina hodgarti*

Green Long Nose Shrimp

The Green Long Nose Shrimp is only rarely found in the trade. It is often imported together with *Caridina* cf. *babaulti* "Green" and is also sold under this name, as neither the collectors in India nor the wholesalers seem to be able to discern between the two species. The elongated rostrum strikes the eye only if you look twice. Some specimens may be of an intense yellow or brown. These shrimp are far more susceptible than other dwarf shrimp from India. In this species, colour changes may take place, depending on their well-being.

Body size: 25 to 30 mm (1 to 1.2")
Temperature: 20 to 31 °C (68 to 88 °F)
pH value: 6.8 to 8.3
Hardness: up to a KH of 25
Tank size: 20 litres (5.2 US gallons) and more
Origins: India
Reproduction: in fresh water

## *Caridina multidentata*

Amano Shrimp, Yamato Shrimp

Its ability to rid an aquarium of unwanted algae makes these shrimp, together with nerite snails of the genus *Vittina*, an ideal first stock in a tank. They do not have any special requirements regarding their environment and do well in almost all aquaria. *Caridina multidentata* comes from the southern part of Central Japan, where it is mostly found in rivers leading to the Pacific Ocean. It also occurs in rivers in Taiwan that lead to the Pacific Ocean.

The females grow much larger than the males. The latter have a dotted pattern on the sides of their pleon, which makes it quite easy to sex these shrimp. The female carries up to 2 000 eggs under its pleon. The larvae need brackish to marine water in order to grow up successfully. In fresh water they will die off after a few days. If you want to raise the larvae you need a separate breeding tank with a salinity of 25 g per litre. The larvae eat Liquizell or similar micro food.

It is really astonishing that these shrimp will live to be eight years old and over, especially if you keep in mind that usually, most dwarf shrimp species only reach an age of two to three years. Amano Shrimp can be co-housed with other shrimp species quite well but can be rather dominant especially when it comes to feeding. Make sure the large, robust Amano shrimp do not leave the smaller shrimp without food.

Body size: 35 to 55 mm (1.4 to 2.2")
Temperature: 16 to 28 °C (61 to 82 °F)
pH value: 7 to 8.3
Hardness: up to a KH of 20
Tank size: 60 litres (16 US gallons) and more
Origins: Japan, Taiwan
Reproduction: larvae need brackish to marine water

## *Caridina serratirostris*

### Ninja Shrimp

Ninja shrimp are very variable in colour and usually occur in the basic colours red, blue, black or brown. Some of them have a longitudinal stripe along their back and horizontal stripes, others, depending on the place they are collected, do not. In stressful situations, they can change their colour all of a sudden. Adult Ninja Shrimp mainly live in the estuaries of rivers, whose salinity fluctuates with the flood, however, some of them are found several kilometres upriver, too. In the aquarium, they can be kept in fresh water all year round.

They are quite shy and usually hide under driftwood and rocks, and they prefer a densely planted tank. The females carry a large number of relatively small eggs, from which planktonic zoea larvae hatch. For their development through several larval stages and moults, they need brackish to marine water to survive, and they return to fresh water only as fully developed shrimp. Up to now, this species has not yet been bred in captivity.

Body size: 20 to 25 mm (0.8 to 1")
Temperature: 20 to 29 °C (68 to 84 °F)
pH value: 6 to 7.5
Hardness: up to a KH of 20
Tank size: 10 litres (2.7 US gallons) and more
Origins: Indonesia, Malaysia, Australia, Madagascar
Reproduction: larvae need brackish to marine water

## *Caridina thambipillai*

Sunkist Shrimp

For a long time, this shrimp was listed under the scientific name *Caridina* cf. *propinqua*. Recent examinations have shown, though, that the Sunkist Shrimp in fact belongs to the species *Caridina thambipillai*, which is found in the brackish estuaries of some rivers in Malaysia and Indonesia. Most Sunkist Shrimp in the hobby are of a stunning orange colour, however, the reason for this phenomenon is a strict selection performed by dealers and exporters in Asia. Depending on their mood and the water they may turn transparent or translucent brown. It is unknown whether these shrimp have already been bred in captivity, as the larvae need brackish to marine water for successful development.

Body size: 20 to 25 mm (0.8 to 1")
Temperature: 20 to 29 °C (68 to 84 °F)
pH value: 6 to 7.5
Hardness: up to a KH of 20
Tank size: 10 litres (2.7 US gallons) and more
Origins: Indonesia, Malaysia
Reproduction: larvae need brackish to marine water

SHRIMP

## *Caridina trifasciata*
Three-Banded Shrimp

German aquarists have brought these shrimp to Europe. The Three-Banded Shrimp is quite variable in colour and pattern, depending on the place it is collected in. It can be totally black, blue-and-white, black-and-white or golden-and-black. Some shrimp are transparent with black stripes. In the tank, the uniform black colour disappears after a short time, fading to a weak blue. It is assumed that the black colour directly depends on plants or the substrate present in the habitat. The water parameters and the overall environment required by this shrimp are similar to that of the Bee Shrimp, which is a part of the species group around *Caridina serrata* as well. The Three-Banded Shrimp does not tolerate temperatures above 25 °C (77 °F) well.

Body size: 20 to 30 mm (0.8 to 1.2")
Temperature: 16 to 24 °C (61 to 75 °F)
pH value: 6 to 7.5
Hardness: up to a KH of 10
Tank size: 20 litres (5.2 US gallons) and more
Origins: Hong Kong
Reproduction: in fresh water

## *Neocaridina davidi*

**Red Cherry Shrimp, Red Fire Shrimp**

The most widely spread shrimp in the hobby is usually called Cherry or Red Cherry, sometimes Red Fire Shrimp. This highly variable species originates from Chinese and Taiwanese waters. They were introduced in Hawaii as fish food, and spread quickly in nature there. Wild-caught shrimp are usually transparent to reddish, some of them have a longitudinal stripe along their back, others don't. Shrimp with transparent parts are called Rili Shrimp.

This species does not have complicated requirements, and it is recommended for beginners. The aquarium size should be chosen well, too small a tank is soon overcrowded, as *Neocaridina davidi* is a highly productive species. No heater is required, and the shrimp are not very demanding when it comes to water parameters.

Body size: 20 to 30 mm (0.8 to 1.2")
Temperature: 4 to 28 °C (39 to 82 °F)
pH value: 6 to 8
Hardness: up to a KH of 15
Tank size: 20 litres (5.2 US gallons) and more
Origins: China, Taiwan, Korea, Japan
Reproduction: in fresh water

## *Neocaridina davidi* var. "Sakura"

### Sakura Shrimp, Red Sakura

The Sakura Shrimp is a deep red colour variant of the normal Red Cherry Shrimp. It was probably first bred from normal *N. davidi* at a shrimp farm in Taiwan. Among its offspring, some transparent or weakly coloured shrimp may crop up. In order to avoid larger numbers of transparent offspring you ought to refrain from socialising Sakura with normal *N. davidi*. Sometimes, newly bought shrimp may lose their colour. Give them a few days to regain their bright, intensive red.

Body size: 20 to 30 mm (0.8 to 1.2")
Temperature: 4 to 28 °C (39 to 82 °F)
pH value: 6 to 8
Hardness: up to a KH of 15
Tank size: 20 litres (5.2 US gallons) and more
Origins: tank-bred form from Taiwan
Reproduction: in fresh water

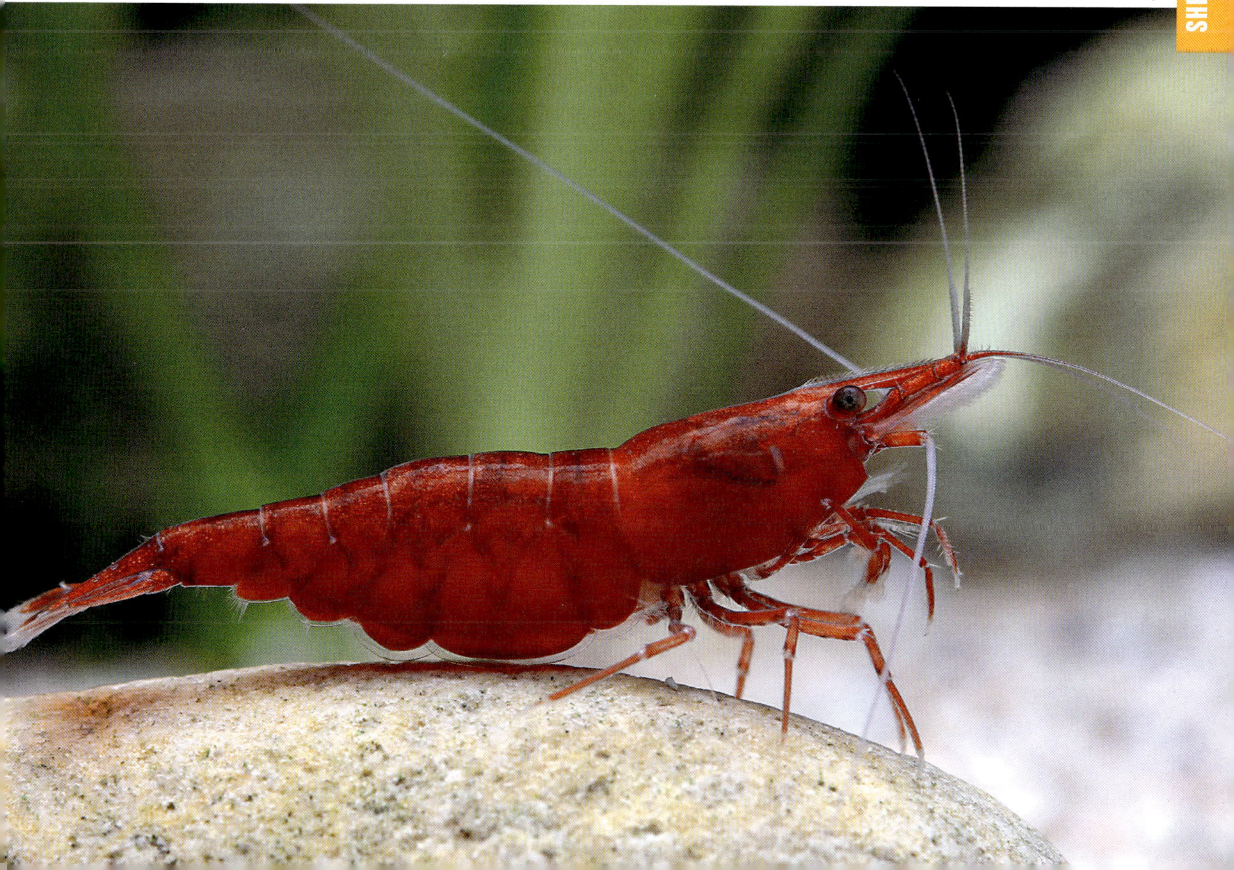

## *Neocaridina davidi* var. "Yellow"

**Yellow Shrimp**

Up until now it is not quite clear whether the yellow variant of *Neocaridina davidi* was created through selective breeding or whether a spontaneous mutation took place. Some Yellow Shrimp have a longitudinal stripe along their back, some don't, just like in the red colour morph. As a rule, adult females are of a more intensive colour, and their eggs are of an intense yellow. Shrimplets tend to be more transparent. This colour morph interbreeds with all other colour variants of *Neocaridina davidi*. If you want their offspring to be exclusively yellow you should keep them separately.

## *Neocaridina davidi* var. "Orange"

Orange Shrimp

The orange colour variant of *Neocaridina davidi* was originally bred in Germany, by selective breeding of a Yellow Shrimp strain. In Taiwan, breeders also worked on an orange strain, which is much more intensive in colour, but whose exact ancestry remains undisclosed. Adult shrimp are light to deep orange, with the females showing a much more intensive colour than the males. Young shrimp are usually of a transparent yellow and turn more orange as they mature. This colour morph interbreeds with all other colour variants of *Neocaridina davidi*. If you want their offspring to be exclusively orange you should keep them separately.

***Neocaridina davidi* var. "Black Fire"**

Black Sakura Shrimp

The Black Sakura strain was bred true in Taiwan, too. Some black specimens were found in nature, too, in a creek in Japan.

## *Neocaridina davidi* var. "Chocolate"

Chocolate Shrimp, Brown Shrimp

As *N. davidi* reproduces readily and easily it is highly probable that we will see more new colour variants and patterns in the future. One example is the new colour morph "Chocolate" originating from Taiwan.

## *Neocaridina davidi* var. "Blue Fire"

Blue Sakura Shrimp, Blue Fire Shrimp, Blue Cherry Shrimp

Only a few specimens of the Blue Sakura are known. There are shrimp with and shrimp without a longitudinal lighter-coloured stripe along their back.

### *Neocaridina davidi* var. "Red Rili"
**Rili Shrimp**

This colour variant comes from Asia. It was originally called Two-Coloured Sakura, but is now known as Rili Shrimp. These shrimp have been bred from discoloured red *Neocaridina davidi* lacking colour in some places.

### *Neocaridina davidi* var. "Yellow Rili"

**Yellow Rili Shrimp**

As in Red Rili, this colour pattern has also appeared in the Yellow Shrimp. Patterns and colour density may vary strongly. Up until now, we do not have a classification system for Rili Shrimp.

***Neocaridina davidi* var. "Orange Rili"**
Orange Rili Shrimp

This variant comes from Germany. At the moment, breeders are working on new special colour patterns, which are not breeding true yet, unfortunately.

### *Neocaridina davidi* var. "Blue Jelly"

Blue Jelly Shrimp

The Blue Jelly Shrimp has been selectively bred from Red Rili Shrimp with a blueish body colour, whose red parts have disappeared bit by bit. Among "normal" Red Rili, you find blueish ones from time to time, and even some without any red pigments.

### *Neocaridina davidi* var. "Blue Rili"

Blue Rili Shrimp

This colour morph could be called the "missing link" between Red Rili and Blue Jelly. It has a blue body colour and still some red parts. With constant sorting according to colour variants, the colour morphs can be stabilised.

## *Neocaridina palmata* var. "White Pearl"

White Pearl Shrimp, Snowball Shrimp

The White Pearl Shrimp is distributed throughout China. Its wild form is of a rather inconspicuous grey to transparent and has all but disappeared from the aquarium hobby. The white colour morph is pretty widely spread in the hobby though. It was first bred in Germany. The shrimp reproduce readily, and they are ideal for beginners. Berried females carry white eggs, which gave them their trivial name. These eggs can be seen while still in the gonads in the back of the neck of the female shrimp (the saddle) before they wander down to the swimmerets where they are attached. The scientific name of this shrimp was *Neocaridina* cf. *zhangjiajiensis* for years, referring to the national park in China of that name where it was first discovered.

Body size: 25 to 35 mm (1 to 1.4")
Temperature: 10 to 30 °C (50 to 86 °F)
pH value: 6 to 8.3
Hardness: up to a KH of 15
Tank size: 20 litres (5.2 US gallons) and more
Origins: tank-bred form from Germany
Reproduction: in fresh water

## *Neocaridina palmata* var. "Blue Pearl"

Blue Pearl Shrimp

As the name tells us, this is a blue variant of the species *Neocaridina palmata*. Its icy blue colour is very attractive. The well-being of the shrimp seems to have an immediate effect on its colour. If you transfer them from one tank to another, they may regain their full colour only after a few days, or they may even stay transparent and colourless for the rest of their lives. A dark substrate seems to be beneficial as they show their nicest colours on dark ground. Like the White Pearl Shrimp, the Blue Pearl Shrimp is recommended for beginners, and it is as productive as the white variant.

Body size: 25 to 35 mm (1 to 1.4")
Temperature: 10 to 30 °C (50 to 86 °F)
pH value: 6 to 8.3
Hardness: up to a KH of 15
Tank size: 20 litres (5.2 US gallons) and more
Origins: tank-bred form from Germany
Reproduction: in fresh water

SHRIMP

## *Paracaridina* sp. "Blue Bee"

Blue Bee Shrimp, Chocolate Bee Shrimp

The shrimp imported under the name Blue Bee or Chocolate Bee belong to a still unidentified species and originate from southern China. In the aquarium they prefer a life in hiding, and their special pattern makes them hard to see – and hard to find in the tank. These shrimp reproduce in fresh water and have been bred in captivity, however, experience shows that they tend to have much less offspring than other dwarf shrimp. The females are usually of a more intense colour and grow a bit larger than the males.

Body size: 20 to 22 mm (0.8 to 0.9")
Temperature: 20 to 24 °C (68 to 75 °F)
pH value: 6 to 8
Hardness: up to a KH of 15
Tank size: 10 litres (2.6 US gallons) and more
Origins: southern China
Reproduction: in fresh water

## *Paracaridina haivanensis*

Princess Bee Shrimp

The species sold under the trivial name Princess Bee Shrimp has been imported from Vietnam by German aquarists quite recently, and it looks a lot like the Bee Shrimp. These shrimp are quite variable in colour and pattern. Besides shrimp with black-and-white bands, there are some with very individual patterns. They can be kept in the aquarium without any problems, and in soft water they reproduce readily.

Body size: 18 to 21 mm (0.7 to 0.8")
Temperature: 17 to 24 °C (62 to 75 °F)
pH value: 6 to 7.5
Hardness: up to a KH of 3
Tank size: 20 litres (5.2 US gallons) and more
Origins: Vietnam
Reproduction: in fresh water

# SHADOW BEE SHRIMP

# Taiwan Bee Shrimp or Shadow Shrimp
# A new generation

New colour morphs originating from Taiwan have caused a lot of excitement in the shrimp scene during the last years. The breeders at first gave them imaginative names like Panda Bee, King Kong, Blue Bolt, Black Diamond, Red Amber or Red Ruby. All of these shrimp are known as Taiwan Bee Shrimp in Europe. In Asia and among breeders in all the world, they are called Shadow Shrimp, Shadow Bee Shrimp or Shadow Bees. Meanwhile, for better orientation and description, a classification system has been established for these colour morphs.

The arrival of the first Shadow Bee Shrimp in Japan and Europe triggered a major hype around these beautiful shrimp in specialised forums and with breeders, quite similar to the interest the shrimp from the Ancient Lakes of Sulawesi had stirred only a short time before.

There are several theories how the Shadow Bee has come into existence. Some sources say the new colour morphs were probably created through a spontaneous mutation that is said to have occurred in the tank of a breeder of the highly variable Red Bee Shrimp in Taiwan. A comparable development was observed earlier on in Germany, however, the outcome was not as attractive. Another theory says that Shadow Shrimp are a result of a cross between Tiger and Bee Shrimp. In our opinion, this theory is viable, as some German breeders have managed to achieve good results from deliberately crossing Tiger and Bee Shrimp (Red Bee Shrimp).

In the beginning, breeders did not follow the most obvious course of only breeding Shadow Bees together, as would be the layman's way to go. The first specimens were crossbred with standard Red Bee and other Bee Shrimp variants and thus, a lot of lines of mixed heritage were created. In the opinion of the breeders, this led to more robust, less susceptible offspring. By now, many stable lines of Shadow Bee Shrimp have been created, and they form the basis for a new generation of beautiful shrimp. New colour morphs are created from this new shrimp generation at astonishing speed.

For breeding Shadow Bee Shrimp, water parameters of 200 µS/cm and a pH of 6.4 to 6.9 have proven suitable. Temperatures ought to be in the range of 22 to 24 °C (72 to 75 °F) for successful breeding. The shrimp can be kept at a greater temperature range of 17 and 26 °C (63 to 79 °F), even with temporary variations, just like their ancestors were used to in their natural habitats. Breeding tanks are, as a rule, planted sparingly, with mosses or undemanding plants like *Vallisneria*.

## Shadow Bee Grades

Black or Red Shadow lined
Black or Red Shadow banded
Black or Red Shadow saddled
Black or Red Shadow Hinomaru
Full Black or Red Shadow
Black or Red Shadow Mosura
Black or Red Shadow Mosura Spotted Head (aka Pinto)
Black or Red Shadow Multistripe
Blue Bolt
Red Bolt

Black Shadow single saddled

Black Shadow triple banded

Black Shadow triple banded

Red Shadow double saddled

Blue Bolt

Black Shadow Mosura Spotted Head

Blue Bolt

Black Shadow Hinomaru

# SULAWESI

Among the shrimp originating from the Indonesian island of Sulawesi, some especially beautifully coloured and strikingly patterned species are found, which rather the eye-catching shrimp we find in marine coral reefs. As they are quite small and have very special requirements they are not suitable for keeping in a mixed tank but ideal for a dedicated nano tank.

## A dedicated Sulawesi tank

During the last few years, many different shrimp species have been imported for the aquarium hobby from the Indonesian island of Sulawesi. Some of the great lakes of Sulawesi are ancient, geologically speaking, and thus form extraordinary biotopes with highly specialised life forms and a large number of endemic shrimp species. Up until now, 21 different shrimp species have been described from the Central Lakes of Sulawesi. Six of them are found in Lake Poso (with a surface area of 320 km²/123 square miles and an average depth of 450 metres/1 476 ft), and fifteen in the Malili Lake System consisting of lakes Matano, Towuti, Mahalona, Lontoa and Masapi. All of these shrimp belong to the genus *Caridina*. They live in different biological niches in the lakes. Some of them have very special demands regarding their habitat, like e. g. *Caridina spongicola*, which only lives in the cavities and on the surface of freshwater sponges.

### Water parameters

Sulawesi shrimp need different water parameters to all other shrimp. These relatively small shrimp only grow to 1.2 to 1.5 cm in length (0.47 to 0.59"). The surface temperature of the lakes is 28 to 32 °C (82 to 87 °F), and the shrimp need high temperatures of 28 to 30 °C (82 to 86 °F) in their aquarium all year round, too. At a depth of 10 to 15 metres, the lake water is still 27 °C (81 °F). Many other water parameters are quite out of the common, too, as the Malili Lake System has a relatively high pH of 7.8 to 8.5, yet the water is quite soft.

### Reproduction

The entire reproduction cycle of the shrimp species from the Ancient Lakes takes place in fresh water, and they can thus be successfully bred in aquaria. All these shrimp have very large eggs in comparison with their body length, however, the females tend to carry relatively small numbers per batch (15 to 20). The eggs are carried under the pleon for about three weeks. The larvae go through all stages of metamorphosis inside the egg shell, there are no benthic larval stages. Once they hatch and leave their mother, the shrimplets are a perfect miniature version of their parents and live on their own from the first second on. In the beginning, they are minute. Thus Sulawesi shrimp should not be socialised with fish, not even with species that have a small adult size, as the minuscule shrimplets will practically fit into any fish mouth, even the tiniest.

### Food

in their natural habitats, Sulawesi shrimp mainly eat biofilms, which they diligently and untiringly graze off all kinds of substrate. In their biotopes, these shrimp live in areas with rocky or stony ground and little to no plants. In the aquarium you can watch them grazing on aufwuchs, too. However, it makes a lot of sense to give them food in addition, e. g. different powdered or flake foods, or tiny plankton (up to the size of a copepod). Their substrate-bound feeding habit makes Sulawesi shrimp act rather uninterested when food is put into the tank, and they do not react as fast and as greedily to artificial foodstuffs as other shrimp species.

### Keeping and socialisation

The first wild-caught Sulawesi shrimp were – and still are – highly susceptible towards changes in their environment, however, the very first tank-bred specimens could be kept at slightly different water parameters, and today some of the more robust species even breed readily at these parameters. However, their temperature acceptance range is still very small, and they need very warm water that should at no time fall below 28 °C (82 °F). Sulawesi shrimp react strongly to disturbances and go into hiding immediately, seemingly leaving the tank shrimpless. They should never be co-housed with fish. Sulawesi shrimp can be kept in nano tanks very well. A group consisting of shrimp of one single species per tank is considered ideal. They can be co-housed with the fascinating snail species that live in their habitats in nature without any problems.

## *Caridina caerulea*

Blue Leg Poso Shrimp

Usually, the Blue Leg Poso Shrimp is transparent with red antennae. The rostrum, the walking legs and the tail fan are blue. However, there are a few all-blue specimens, too. The Blue Leg Poso Shrimp lives on rock, wood or stones in the habitat, and sometimes it is also found on soft substrates like macrophytes (aquatic plants).

The species is endemic in Lake Poso. It is the only Sulawesi species with blue pigmentation. Most other species are red or brown.

Body size: 22 mm (0.9")
Temperature: 28 to 30 °C (82 to 86 °F)
pH value: 7.8 to 8.5
Hardness: KH of 5 to 7
Tank size: 20 litres (5.2 US gallons) and more
Origins: Sulawesi, Lake Poso
Reproduction: in fresh water

## *Caridina dennerli*

### Freshwater Cardinal Shrimp

*Caridina dennerli* is one of the most beautiful shrimp from Sulawesi. Like *Caridina loehae*, the Freshwater Cardinal Shrimp is found between or under rocks. It lives in shallow zones close to the lakeshore, but can sometimes be found at depths of down to 10 metres. Unlike most other known shrimp species, the Freshwater Cardinal Shrimp has very few shrimplets per batch. The females carry only about six to fourteen eggs, thus an overpopulation in the aquarium is rather improbable. Besides red shrimp with white dots there are also some with blue dots or even larger blue spots. If they are under stress they can turn temporarily blueish. However, they go back to their original colour within minutes.

Often they are quite shy in the aquarium and prefer rock crevices where they sit during the day, just as in nature. There you can see them move their strikingly white chelipeds, with which they untiringly graze detritus and biofilm off the rocks and the substrate. Meanwhile, *Caridina dennerli* has been bred in captivity, and the problems that occurred in the beginning have been overcome.

Body size: 15 mm (0.6")
Temperature: 27 to 30 °C (81 to 86 °F)
pH value: 7.8 to 8.5
Hardness: KH of 5 to 7
Tank size: 20 litres (5.2 US gallons) and more
Origins: Sulawesi, Lake Matano
Reproduction: in fresh water

## *Caridina glaubrechti*

Red Orchid Shrimp, Brown Camo Shrimp

The most striking characteristic of this shrimp is its extremely long, serrated rostrum. Up until now, they have only been imported for the aquarium hobby rarely, and they are not as widely spread as e.g. *Caridina dennerli*. The Red Orchid Shrimp lives on rocks and stones. It is most frequently found under small rocks near the shore or between and under larger boulders in zones of medium depth (down to three metres).

Body size: 14 mm (0.55")
Temperature: 28 to 30 °C (82 to 86 °F)
pH value: 7.8 to 8.5
Hardness: KH of 5 to 7
Tank size: 20 litres (5.2 US gallons) and more
Origins: Sulawesi, western part of Lake Towuti
Reproduction: in fresh water

## *Caridina holthuisi*

Towuti/Matano Tiger Shrimp

SHRIMP

*Caridina holthuisi* got its name in honour of the recently deceased Professor Holthuis, a renowned carcinologist. It is very variable in colour and usually lives on soft substrates, frequently hiding under dead plant parts and detritus. Sometimes it is found on aquatic plants, too. In order to camouflage from predators these shrimp are usually dark brown to black. Some have stripes, and others do not have any patterns at all. Individual specimens may even be white, depending on the habitat they come from and the colour of the substrate.

Body size: 24 mm (0.9")
Temperature: 28 to 30 °C (82 to 86 °F)
pH value: 7.8 to 8.5
Hardness: KH of 5 to 7
Tank size: 20 litres (5.2 US gallons) and more
Origins: Sulawesi, Lakes Matano, Mahalona and Towuti, Petea River
Reproduction: in fresh water

## *Caridina loehae*

Orange Delight Shrimp, Mini Blue Bee

This species lives on hard substrates and is most frequently found under or between rocks near the lakeshore. It is highly variable and can change colours within a very short time when under stress. After transport, Orange Delight Shrimp can be of a purplish blue, but after a short adaptation time they turn red or reddish brown with tiny white dots. Berried females carry reddish to intensely red eggs. As rock dwelling shrimp, *Caridina loehae* inhabits the same biotopes as *Caridina dennerli* and *Caridina parvula*.

Body size: 13 mm (0.5")
Temperature: 28 to 30 °C (82 to 86 °F)
pH value: 7.8 to 8.5
Hardness: KH of 5 to 7
Tank size: 20 litres (5.2 US gallons) and more
Origins: Sulawesi, Lakes Matano and Towuti
Reproduction: in fresh water

## *Caridina profundicola*

Towuti Sun Ray Shrimp

Together with *Caridina spinata*, this beautiful shrimp is found in the deeper regions of Lake Towuti. It usually lives in waters over three metres deep, where it occurs under larger rocks. The striking yellow stripes on its body distinguish it from all other species in the lake. Its bright green eggs contrast nicely against its yellowish body colour.

Body size: 16 mm (0.6")
Temperature: 28 to 30 °C (82 to 86 °F)
pH value: 7.8 to 8.5
Hardness: KH of 5 to 7
Tank size: 20 litres (5.2 US gallons) and more
Origins: Sulawesi, Lake Towuti
Reproduction: in fresh water

## *Caridina spinata*
Red Goldflake Shrimp

*Caridina spinata* is one of those shrimp preferring to live under and between rocks or on pieces of wood or fallen trees. You can find these shrimp at several places in the lake, where they occur from shallow areas near the shore down to a depth of three to five metres. Adult shrimp seem to go to the deeper zones more frequently, where they search the ground for edible stuff together with other rock-dwelling species like *Caridina profundicola*.

The Red Goldflake Shrimp is variable in pattern and colour. It may be of a deep dark red with orange or yellow stripes or dots, but its colour can vary to an almost unicoloured orange. It is one of the larger shrimp species of Lake Towuti, and due to its colour it can easily be discerned from other shrimp species that share its biotope.

Body size: 15 mm (0.6")
Temperature: 27 to 30 °C (81 to 86 °F)
pH value: 7.8 to 8.5
Hardness: KH of 5 to 7
Tank size: 20 litres (5.2 US gallons) and more
Origins: Sulawesi, Lake Towuti
Reproduction: in fresh water

## *Caridina striata*

Red Line Shrimp

*Caridina striata*, a rock-dwelling species, is usually found under stones near the lakeshore, however, it also occurs in zones with a depth of up to three metres. Its large morphological variability is its most striking characteristic. The serrate rostrum may vary in length and in form, from a sabre-shaped upward curve to almost straight. Colour and pattern do not vary that much, the shrimp all have thin longitudinal red and white stripes.

Body size: 14 mm (0.5")
Temperature: 27 to 30 °C (81 to 86 °F)
pH value: 7.8 to 8.5
Hardness: KH of 5 to 7
Tank size: 20 litres (5.2 US gallons) and more
Origins: Sulawesi, Lakes Mahalona and Towuti
Reproduction: in fresh water

## *Caridina woltereckae*

Freshwater Harlequin Shrimp

It was one of the first shrimp exported from Sulawesi for the aquarium hobby in larger numbers, and it was offered at horrendous prices in the beginning. *Caridina woltereckae* lives, just like many other species of the genus *Caridina*, on or under stones and in rock crevices. It is not only found in areas near the lakeshore but also in medium-deep zones down to a depth of three metres, where it inhabits the same ecological niches as species like *Caridina spinata*, *Caridina striata* and *Caridina glaubrechti*. Its typical striped pattern sometimes leads to confusion with *Caridina spongicola*, a shrimp only found in a single spot in Lake Towuti, living exclusively in and on freshwater sponges, whereas *Caridina woltereckae* prefers hard substrates.

Body size: 17 mm (0.7")
Temperature: 27 to 30 °C (81 to 86 °F)
pH value: 7.8 to 8.5
Hardness: KH of 5 to 7
Tank size: 20 litres (5.2 US gallons) and more
Origins: Sulawesi, Lake Towuti
Reproduction: in fresh water
Atya gabonensis

## FAN SHRIMP

Atyid fan shrimp can be subdivided into five genera and around twenty species. However, only three genera have found their way into the aquarium hobby. All fan shrimp have long bristles on their first and second leg pairs that look a lot like Chinese hand-held fans, with which they filter edible particles from the current in fast-flowing waters.

### *Atya gabonensis*

**Giant Filter Shrimp, African Filter Shrimp, Vampire Shrimp**

The Giant Filter Shrimp inhabits flowing waters close to the coast. It prefers rivers with a steep gradient and a correspondingly strong current. Usually, these shrimp are blue or blueish-grey, however, there are white, green or pink specimens and all shades in between. Especially young Giant Filter Shrimp are rather variable in colour. When new in the tank they tend to be very shy and go into hiding at once, where they usually stay for several days in a row. In the habitat, these shrimp only leave their hiding-places at night, when the large fish that predate on them are asleep. When they have adapted to their new surroundings they will lose their shyness and will come out during the day, too. Even though they reach an impressive size they are absolutely harmless for other aquarium inhabitants as they eat floating debris exclusively. A strong current is thus absolutely vital. Breeding in captivity has been successful in only a few cases so far, and it is impossible in a freshwater tank. The larvae need marine conditions for their development, and they return to fresh water only when fully developed into a shrimp, which takes several weeks.

Body size: 100 to 140 mm (3.9 to 5.5")
Temperature: 20 to 30 °C (68 to 86 °F)
pH value: 6.25 to 7.5
Hardness: up to a KH of 10
Tank size: 50 litres (13 US gallons) and more
Origins: Western Africa, east coast of South America
Reproduction: larvae need marine water

## *Atyopsis moluccensis*

Bamboo Shrimp, Wood Shrimp

Like most other fan shrimp, Wood Shrimp prefer rapidly flowing waters, where they sit in the strong current between and on rocks, clinging onto the surface, filtering small particles and floating debris from the water. This species is quite variable in colour and can be from dark brown over green to yellowish-orange. Depending on the area of collection they also vary in pattern. Wood Shrimp from Sri Lanka have a dark striped pattern, those from Malaysia dark spots, and those from the Philippines have a light-coloured stripe along their back.

In the aquarium, they need a very good current. They have rarely been bred in captivity as the larvae need marine conditions for successful development. The Wood Shrimp prefers lowland rivers to mountain creeks.

Body size: 80 to 90 mm (3.1 to 3.5")
Temperature: 24 to 28 °C (75 to 82 °F)
pH value: 6.5 to 7.5
Hardness: up to a KH of 15
Tank size: 60 litres (15 US gallons) and more
Origins: Indo-Pacific region
Reproduction: larvae need marine water

## *Atyoida pilipes*

Green Lace Shrimp

The Green Lace Shrimp is found in small, rapidly flowing creeks leading to the sea. After a time needed for adaptation these at first rather nocturnal shrimp lose their shyness, and you can see them during the day, too, as they filter food particles from the current, putting them into their mouth with their fans. As with all fan shrimp, their larvae need marine conditions to develop, requiring a longer time in the sea until they metamorphose into adult shrimp.

Even though this fan shrimp species does not grow very large the tank ought to have at least 60 litres (15 US gallons). Small groups of three to five Green Lace Shrimp are considered ideal. If the water level in the tank is very high it may happen that they leave the tank. In nature, they have been observed to climb rocks and to go up small waterfalls. Please make sure your Green Lace Shrimp tank is tightly covered.

Body size: 40 to 60 mm (1.6 to 2.4")
Temperature: 22 to 26 °C (72 to 79 °F)
pH value: 7.5 to 8
Hardness: up to a KH of 15
Tank size: 60 litres (15 US gallons) and more
Origins: South-East Asia: Philippines, New Guinea, some Pacific islands
Reproduction: larvae need marine water

## LONG-ARM SHRIMP

Long-arm Shrimp belong to the superfamily Palaemonoidae. They differ from all other shrimp species as their second pair of walking legs is considerably larger. Sometimes, especially in male specimens, they can be even longer than the body of the shrimp itself. They are used for catching food, and, besides that, for defence. Long-arm shrimp are often mistaken for river crayfish, as in some species, the elongated chelipeds are thicker, too.

## *Macrobrachium dayanum*
Red Rusty Shrimp, Redclaw Shrimp

As in all other species of the genus *Macrobrachium*, the males of the Red Rusty Shrimp are stronger, have larger claws and often possess a different pattern and colour from that of the females. The second pair of claws is considerably larger, especially in dominant males. Usually, there is only one male per population that has the large claws as secondary sex characteristics, in the rest of the males this development is suppressed. For this reason, they resemble females very much and thus are not constantly under attack by the aggressive alpha male, especially in the rather confined surroundings of an aquarium.

For reproduction, merely the most elementary of requirements need to be fulfilled. If the tank is overstocked or they are not fed sufficient amounts of protein, the adults as well as even young *Macrobrachium dayanum* tend to cannibalise each other. Young Red Rusty Shrimp have circular red patterns on the legs, the most eye-catching ones around their chelipeds, which fade with age. Adult shrimp are, as a rule, of a uniform light to dark brown.

Body size: 70 to 90 mm (2.8 to 3.5")
Temperature: 14 to 28 °C (57 to 82 °F)
pH value: 7 to 8.3
Hardness: up to a KH of 20
Tank size: 50 litres (13 US gallons) and more
Origins: India and Bangladesh
Reproduction: fresh water

## *Macrobrachium lanchesteri*

Ghost Shrimp, Glass Shrimp

Even though the Ghost Shrimp is not a very demanding species, not many reports of them breeding successfully in captivity were known in the beginning. Even though they may reproduce on a regular basis, when space is limited, the parents predate heavily on their offspring, and the offspring also cannibalise each other. However, they are quite harmless for most other tank inhabitants, especially fish, as they stay relatively small for a long-arm shrimp. As a rule, they do not attack fish at all and can thus be kept in a social fish tank, given that the fish are not too large or predatory.

The Ghost Shrimp is more susceptible towards lower temperatures than the Red Rusty Shrimp. In this species, physical differences between dominant and subdominant males are common, too, even though their claws are not as large and as strong as those of other long-arm shrimp.

Body size: 50 to 60 mm (1.9 to 2.3")
Temperature: 23 to 32 °C (57 to 89 °F)
pH value: 6 to 8.3
Hardness: KH of 15 to 30
Tank size: 50 litres (13 US gallons) and more
Origins: Burma and Thailand
Reproduction: fresh water

## *Macrobrachium idae*

Long-Arm Shrimp

The body colour of these shrimp is yellowish-brown and slightly marbled with a darker hue. Especially striking are its extremely long chelipeds. Those of the males in particular are truly an eye-catcher, as they are also considerably thicker. However, the females may develop into nimble fish catchers with their more slender legs as well, as their reach is also impressive and they are nifty and fast with their pincers.

Any and all fish socialised with these long-arm shrimp will turn into shrimp food sooner or later, thus they should not be kept together. These shrimp are nocturnal and hardly ever show in too bright a tank, thus the lighting should be kept subdued. Floating plants can be used to dim the tank, too. *Macrobrachium idae* does not tolerate low temperatures, it prefers warmer water as it comes from the Indian and Indo-Pacific tropics. Its eggs are relatively small compared to its body size, and the larvae only develop in marine conditions.

Body size: 100 to 110 mm (3.9 to 4.3")
Temperature: 20 to 29 °C (68 to 84 °F)
pH value: 7.5 to 8
Hardness: up to a KH of 15
Tank size: 120 litres (32 US gallons) and more
Origins: East Africa and Madagascar, South-East Asia to Australia
Reproduction: larvae need marine conditions

## *Macrobrachium rosenbergii*
Giant Malaysian Prawn

The Giant Malaysian Prawn grows extremely large. It occurs in stagnant and flowing waters near the coast, and due to its gigantic size it plays an important role in shrimp farming, too. These exceptional shrimp are beautifully coloured. Their body is brownish to greyish blue, and their chelipeds are of a deep, attractive electric blue. However, their considerable size and their partially predatory lifestyle make them of limited use for a common aquarium. Adult males may grow to a length of over 50 cm (20") counting their chelipeds, and need a correspondingly large tank. They are not a peaceful species either. You cannot keep more than one male together with a few females even in a very large tank. The Giant Malaysian Prawn is mainly nocturnal and should never be kept in a social tank. Hardly any tank inhabitant – maybe with the exception of brown plecos – is bound to survive when kept together with these shrimp.

Body size: 250 to 320 mm (9.8 to 12.6")
Temperature: 22 to 29 °C (72 to 84 °F)
pH value: 7.5 to 8.5
Hardness: KH of 15 to 20
Tank size: 500 litres (132 US gallons) and more
Origins: South-East Asia to Australia
Reproduction: larvae need brackish to marine conditions

## *Arachnochium kulsiense*
Pearl Shrimp

*Arachnochium kulsiense* was scientifically described only recently. It originates from a tributary of Brahmaputra River in India. It stays comparatively small, and even adult males only have thin, short chelipeds. Its body is of a transparent white and has innumerable white, brown and blueish-black dots. Its legs and pincers have pale blue patterns and rings.

In contrast to many species of the genus *Macrobrachium*, the Pearl Shrimp is very peaceful towards conspecifics and can thus be kept in larger groups. If you do, however, please make sure to keep them in very well-structured tanks that provide them with lots of places to hide as they lead a rather secluded life. If they cannot satisfy their need to take cover they will become very shy and visibly unwell. If they are comfortable they will reproduce in the tank. This is possible in pure fresh water given that the water is sufficiently hard. However, the Pearl Shrimp is not a very productive shrimp. A female has hardly more than 20 eggs per batch. Populations explosions are therefore practically ruled out.

Body size: 40 to 60 mm (1.6 to 2.3")
Temperature: 20 to 26 °C (68 to 79 °F)
pH value: 7.5 to 8.3
Hardness: KH of 10 to 20
Tank size: 50 litres (13 US gallons) and more
Origins: northern India
Reproduction: fresh water

## *Macrobrachium eriocheirum*

Fuzzy Claw Shrimp

*Macrobrachium eriocheirum* lives in flowing waters of different width. Its colour varies from yellowish to reddish brown and changes depending on the environment and especially the substrate colour. Like all representatives of this species group, male Fuzzy Claw Shrimp have visibly enlarged claws on their second walking leg pair. One of the claws is much thicker than the other. All eleven species of the group have strikingly long bristles on their claws. The dactyl and the propodus segments (the "hand" and the "finger") look as if they are covered in grey velvet, from which longer bristles protrude.

They are rather undemanding shrimp and will soon adapt to a well-decorated tank where they can move vertically as well as along the walls. They do not damage aquatic plants, however, other tank inhabitants may be in danger, as they tend to predate on smaller fish and inverts. As long as its most basic requirements are fulfilled, the Fuzzy Claw Shrimp will reproduce readily.

Body size: 60 to 80 mm (2.3 to 3.1")
Temperature: 20 to 26 °C (68 to 79 °F)
pH value: 7 to 7.5
Hardness: up to a KH of 15
Tank size: 50 litres (13 US gallons) and more
Origins: China to Thailand
Reproduction: fresh water

## *Macrobrachium hendersoni*

Henderson's Shrimp, Red Claw Shrimp

Even though *Macrobrachium hendersoni* belongs to the same species group as *Macrobrachium assamense assamense* and *Macrobrachium dayanum* it grows a bit larger than these two species. The three species share their area of distribution, too. Young shrimp of these closely related species are very hard to discern, as they all have bright red rings around their legs. They are greyish brown and usually have a reddish brown rostrum. In adult shrimp, the red rings fade almost entirely, only some female specimens still show very faint remainders of this pattern. Like other long-arm shrimp, *M. hendersoni* need a very well-structured tank that provides them with a sufficient number of caves. They breed readily in fresh water. There is only a single larval stage – an adaptation to life in fresh water only. In a social tank you need to be careful – sometimes everything goes great, and sometimes the shrimp will eat fish and other tank inhabitants. When abundantly fed they will not be overly cannibalistic.

Body size: 70 to 80 mm (2.7 to 3.1")
Temperature: 20 to 27 °C (68 to 81 °F)
pH value: 6 to 8
Hardness: up to a KH of 20
Tank size: 80 litres (21 US gallons) and more
Origins: India to Thailand
Reproduction: fresh water

## *Macrobrachium assamense assamense*

Red Claw Shrimp

*M. assamense assamense* is often mistaken for *M. hendersoni*, and indeed the two species are very hard to discern. However, a typical characteristic of *M. assamense assamense* are grooves in the dactyls of the claws. Moreover, the claws of adult males are densely covered with short, thin bristles of a greyish-white colour, which stand out clearly against the otherwise brown chelipeds. The second cheliped pair is much thicker in males, but not significantly longer. The claw on the first cheliped pair is only used for bringing food to the mouth and for cleaning. If you want to keep several of these shrimp in one tank and plan on breeding them, too, make sure the aquarium is really well-structured. As these shrimp do not harm plants they can be kept in a well-planted tank. However, they also do well in completely unplanted tanks as long as they find a sufficient number of caves and other places to hide in structures of rocks and wood. In nature, this species occurs in flowing as well as stagnant waters. It is rather undemanding when it comes to water parameters, however, it still needs clean water and water changes at regular intervals to stay healthy. *M. assamense assamense* is a species that reproduces easily and under various conditions. Young shrimp have red and black bands on their chelipeds, which has given them their name. These patterns fade when they get older, and adult shrimp usually have unicoloured brown legs with a very faint banded pattern.

If they are fed with unsuitable food and if the tank is too small this rather peaceful species may start eating conspecifics and sometimes even fish that don't pay attention. Should you notice partially eaten tail fins of fish in a tank with these shrimp it's time to increase the protein content of the shrimp food.

Body size: 60 to 70 mm (2.3 to 2.8")
Temperature: 14 to 28 °C (57 to 82 °F)
pH value: 7 to 8.3
Hardness: up to a KH of 20
Tank size: 50 litres (13 US gallons) and more
Origins: Burma to India
Reproduction: fresh water

## *Euryrhynchus amazoniensis*

Peru Zebra Shrimp

Body size: 18 to 20 mm (0.7 to 0.8")
Temperature: 25 to 30 °C (77 to 86 °F)
pH value: 4.5 to 6.5
Hardness: up to a KH of 8
Tank size: 20 litres (5 US gallons) and more
Origins: Brazil
Reproduction: fresh water

Unfortunately, this small long-arm shrimp is hard to come by in the trade, even though it is actually a highly suitable shrimp for the aquarium hobby and would merit a wider distribution. Other than its name suggests it comes from small tributaries in the Amazon region, where it lives in large clumps of fallen leaves. It prefers a nocturnal life, which is understandable as it is not a very big shrimp and there are many predators in these waters. Up until now, four different species have been described, however, they cannot be identified by the untrained eye. Their requirements are very similar anyway. They need a dedicated tank. In a social tank they will only survive when kept together with very small fish.

The claws are quite strong in comparison with the rest of its body, and look rather like the claws of a crayfish. These shrimp reproduce in fresh water. They only have a single larval stage, in perfect adaptation to their habitat. Unfortunately, the Peru Zebra Shrimp is not very productive. The females only carry 10 to 15 eggs per batch. This is one of the reasons why this shrimp is not readily available in the trade.

# CRAYFISH

# 27 SPECIES CRAYFISH

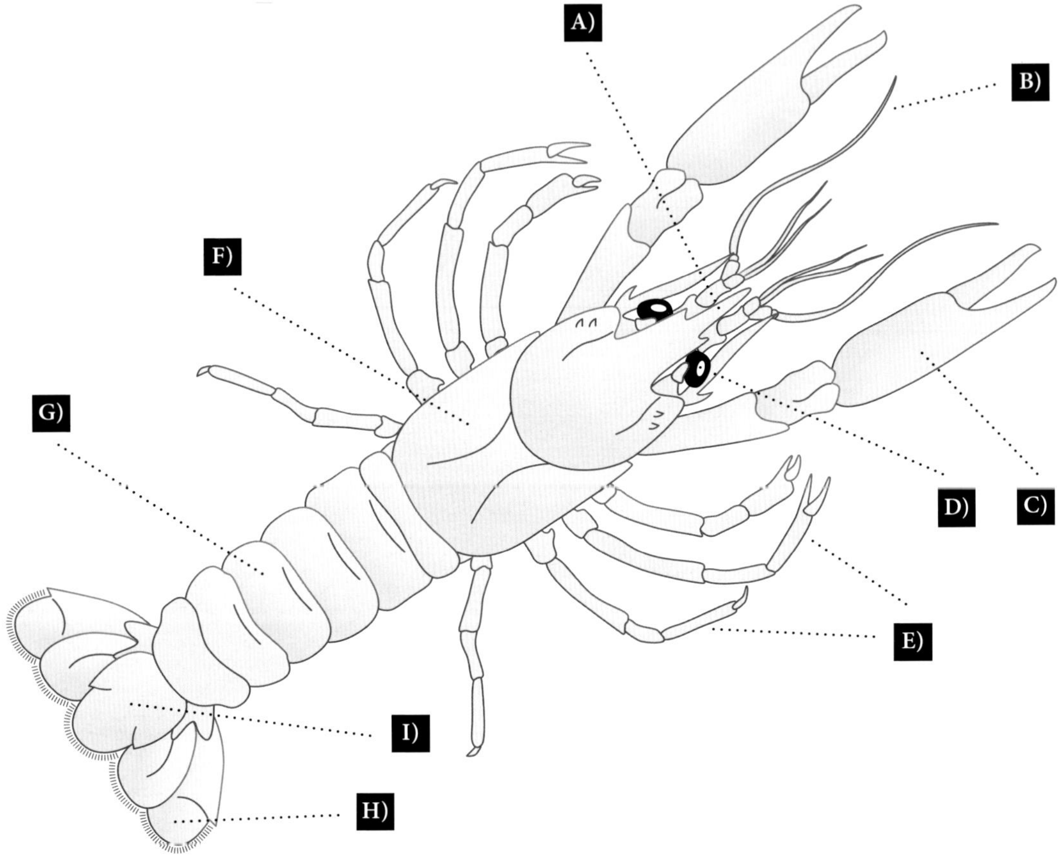


# ANATOMY

A) rostrum
B) antennae
C) claw
D) eye
E) walking leg
F) carapace
G) abdomen/pleon
H) uropods (tail fan)
I) telson

There are approximately 600 crayfish species in the world, and every year, new species are discovered. Crayfish inhabit creeks, streams and lakes, swamps and wet meadows. Some species live in water bodies that fall dry during a part of the year, and some are even completely independent of surface water as they dig tunnels in the ground that are several meters deep and lead to ground water, where the crayfish get all the water they need. On the surface, nothing would make a layman suspect that crayfish can live in such a place, but experts find their burrows because of the chimneys they build, which look a bit like molehills.

Even in cave waters almost devoid of nutrients several species are found. Some of them have been living in these lightless biotopes for so long that they have completely lost their colour and are now of an unpigmented white. In adaptation to life in eternal darkness, in some species the eyes are almost or even entirely degenerated. Some crayfish species have conquered the dry land, and spend the largest part of their lives out of the water. They only return to their original element in order to moisten their gills and to release their offspring. If you consider the wide area of distribution this group of inverts has, the various climate zones and altitudes they inhabit, and take the very different biotopes into regard where crayfish can live you will soon realise that general statements about the requirements of crayfish and the conditions they should be kept in when in captivity are practically of no avail. If you want to create the best artificial habitat possible in your aquarium for your crayfish it is crucial for you to know exactly which species you have. Only then can you be totally sure you meet its individual requirements.

***Procambarus pictus*** **habitat in Florida (USA)**

There are large differences in crayfish body size, too. The bodies of the smallest specimens grow to only one centimetre (0.4") in length, making them veritable dwarves, especially in contrast to the largest crayfish of sixty centimetres (24") in length and up to five kilograms (11 pounds) in weight.

In the aquarium hobby, crayfish have been a late addition. The first ones to end up in an aquarium were originally imported as food for human consumption. Among them, hobbyists found very colourful specimens that kindled their interest. One of the first crayfish in a pet shop tank was the Louisiana Crayfish *(Procambarus clarkii)* from the south of the USA. Starting in the year 2000, a veritable run for crayfish has taken place in the aquarium hobby in Europe, Japan and the USA, and the species pet shops have on offer have become more varied.

All crayfish are bound to fresh water for their entire life cycle. Their embryonic development takes place inside the egg, with the exception of the last one or two stages. In contrast to most marine invertebrates, crayfish do not go through free-floating, planktonic or independent larval stages. The eggs are carried under the pleon of the mother crayfish, hanging from a tiny filament, and the young crayfish that have already hatched are still connected to their mother with this so-called telson thread so they do not get lost. It is fixed to the last body segment of the larvae, their telson, and goes through the egg shell to the swimmerets of the berried female. No other representative of the crustaceans has developed a comparable system.

There is one all-important aspect of crayfish keeping that goes far beyond the aquarium hobby and that needs to be brought to the general public's attention. As global trade with these invertebrates is booming, they have started to become an increasing danger for the environment: Should crayfish be released into nature by irresponsible people they pose a grave danger for the native crayfish population and other aquatic organisms – the entire biotope may be seriously compromised.

**Young *Procambarus alleni***

## The crayfish tank

When the volume is the same, an aquarium with a larger footprint is always preferable, since water depth is not as important for crayfish as the usable surface. It is crucial for the number of crayfish you can keep in a tank. Depending on the size and the aggressiveness of a species, tank decoration is also a very important factor. Crayfish need structure, horizontal as well as vertical, and a sufficient number of places to hide. Moreover, they are veritable escape artists, and they will use every tube, hose and cable leading out of the tank to have a closer look at their surroundings – which, unfortunately, often ends in their death. Even a tad of silicone jutting out at the corner of the tank can form a foothold sufficiently large to enable the crayfish to flee from the aquarium. Thus, a tank cover is not only absolutely necessary, moreover, it needs to be made cray-tight. Make sure you close off every cable feedthrough with foam or netting. If your crayfish belong to a larger species, keep in mind that they have enormous strength and can lift tank covers seemingly without effort if their mind is set on leaving the aquarium.

### Decoration

The majority of crayfish need a cave for their well-being. Some species also take cover in dense vegetation and only build tunnels when their aquatic habitat threatens to fall dry so they need a moist, well-protected retreat. As it is hardly possible to have a loamy stable ground in the aquarium for the crayfish to burrow in we need to provide them with other materials with which they can build their caves. A slab on a sufficiently thick layer of sand and gravel propped up by smaller rocks is the simplest variant. The crays can dig out the gravel below and thus create a tunnel for themselves. However, driftwood, coconut shells, clay flowerpots or clay tubes are suitable hiding-places, too. There are hardly any limits for a designer's imagination. The only important thing is that each crayfish has its own cave to build and to live in. It is even better if the number of hiding-plac-

Clay tubes are ideal hideaways for crayfish.

es exceeds the number of crayfish in the tank so they have the possibility to choose. Many crayfish species are nocturnal or crepuscular creatures, which makes lighting somewhat less important at first glance.

## Plants

Some species do not allow a permanently planted tank as they eat plants right away. Even more robust, bad-tasting plant species are not left without damage, as crayfish tend to walk around in their tank a lot, and burrow, which can harm the plants considerably. Plants floating on the surface and in the water, like Hornwort (*Ceratophyllum demersum*) or Waterweed *(Egeria densa, Elodea canadensis*, which may be illegal for sale and distribution in some countries) are suitable for crayfish tanks, though.

## Filtration

A strong filter is of great advantage in a crayfish tank. They break down their food into very small parts when eating, no matter if they have tablet food or a piece of fish, and small particle clouds tend to spread and spoil the water. Some species may be socialised with small fish or dwarf shrimp with identical requirements regarding the water parameters. These fellow occupants soon learn how to make use of these additional sources of food, and take care of a

Hornwort *(Ceratophyllum demersum)*

large part of these food particles, which would otherwise go to waste.

Their requirements regarding water quality differ from species to species, however, crayfish generally tolerate a rather large range of water parameters. The great robustness and adaptability of some crayfish species should, however, not lead their keepers to the assumption that regular water hygiene is unnecessary. On the contrary, good filtration and regular water changes are a must in every crayfish tank.

## Growth and moulting

The exuvia ought to remain in the tank after a successful moult

Crayfish have a very hard exoskeleton, and thus they cannot grow continuously. When they shed their old skin (or moult) they go through a growth leap. A crayfish is never fully grown, but will continue to shed its skin and to grow immediately after every moult. Of course, they moult more frequently and thus grow faster while still young, and their relative size and weight gain per moult is larger than in older crays. Very old crayfish moult but do not grow significantly at all, they just renew their old skin that is sometimes overgrown with algae and other micro-organisms. Every moult is a life-threateningly dangerous moment for a crayfish. Often, an unsuccessful moult marks the end of a crayfish life.

If the old skin gets too tight, it is shed in order to make room for a new, larger one. For this purpose, the old exoskeleton gets softer and more elastic as hormones trigger metabolic processes that remove calcium from it. After that, the crayfish can slip out of its old skin. This process starts with the skin between the carapace and pleon splitting open. The carapace is lifted upwards towards the front, and the crayfish frees itself, making pumping movements with its entire body. It needs to withdraw its enormously large muscular claws through the narrow tubular leg segments. To enable the crayfish to successfully conclude this process, the leg segments split open longitudinally in order to provide the claws with suffi-

**When moulting, the crayfish frees itself from its old skin with rhythmic movements within a short time. The moulting process typically only takes a few minutes. During this time, the crayfish is in critical danger, as the new skin is still soft and offers no protection from predators. After one or two days, the new skin is completely hardened.**

cient room. The skin on all limbs and even a part of the stomach is shed during a moult.

When the crayfish has managed to free its head and the front part of its body from its old skin and to retrieve its legs and claws from their narrow tubes, it starts flaying its pleon and tail violently to get out of its old frock entirely. After this procedure, the crayfish will be exhausted and, as a rule, sit beside its old shell for a while. Before moulting, some crayfish will climb up plant stems in order to shed their old skin in the oxygen-rich water near the surface and to get away from their conspecifics, which may pose a deadly danger to them during a moult.

**Exuvia of *Cambarellus patzcuarensis***

## Diet and Food

As crayfish inhabit most different biotopes, it is hard to make a general statement about their diet. Usually, crayfish are omnivores, which means they eat food of vegetable as well as of animal origin. Most species will eat anything organic, in fact. The following choice of foodstuffs they eat in nature demonstrates their variability and adaptability regarding food very well: rotting wood, detritus, algae, fungi, bacterial layers, aquatic plants, dead tree leaves, terrestrial plants, fruit, seeds, insects, crustaceans, worms, molluscs, fish and other vertebrates.

Depending on the species, the age of the crayfish and the seasonally available food, its diet will be more on the carnivore or on the herbivore side. For most crayfish, insect larvae and aquatic snails are a favourite when it comes to food of animal origin. They also like fish, even though not many crayfish hunt fish actively but prefer eating recently deceased ones. It is often said that crayfish eat carrion, which is basically untrue, though, as they won't touch anything that has long been dead. They like rotting plants as part of their food, however, they refuse to eat rotting meat. They love a varied diet and are opportunistic feeders, which ought to be kept in mind when it comes to feeding them in an aquarium.

In the trade you will find a large choice of different food, from flakes or pelleted food for fish down to crayfish food especially tailored to their needs. You can widen the range of food by giving grain (maize or wheat work very well, even uncooked), vegetables (carrots, peas, lettuce from organic farming) and fish or mussel meat. A great protein source are aquatic snails, which may live in the tank as a permanent food source. Crayfish are irregular eaters and do not accept any food before they moult, for example. It may thus happen that food remains uneaten. Make sure to remove these remains before they rot away and spoil the water.

Detritus (rotting organic material, especially plant rests) are an inherent part of the diet of a crayfish and should be given to aquarium crayfish, too. For doing so, just pick up dead brown tree leaves from the ground. Rinse them off and put them in your tank. After some time you'll see that crayfish prefer some tree leaves over others.

Oak and beech usually get eaten first. Avoid leaves of strongly aromatic or poisonous trees! Sometimes you'll see your crayfish gnaw on moorwood or other driftwood pieces, which helps them cover their need for fibre. Crayfish offspring will grow up better if they find leaves and detritus in the tank.

Crayfish are not picky eaters but like a varied diet nonetheless.

## Diseases

Crayfish plague is probably the most well-known crayfish disease. It is caused by the water mould *Aphanomyces astaci* from the oomycetes family and it has had a cataclysmic impact on the native crayfish populations of Europe. North-American crayfish have evolved in co-existence with this water mould and have thus been able to develop successful defence mechanisms. In all other species tested up until now, this infection has an incredibly fast progression with an extremely high mortality, entire populations are literally wiped out. In the second half of the 19th century, the pathogen was probably introduced into Europe accidentally, when infected crayfish from North America were imported in great numbers. Since then, massive population wipe-outs have been recorded in nature. Native European crayfish are attacked by the zoospores, which can move with the help of two flagella. Only one day after infection, the host's behaviour changes. Infected crayfish start cleaning and scratching themselves a lot more than usual and show restless movement. However, these ini-

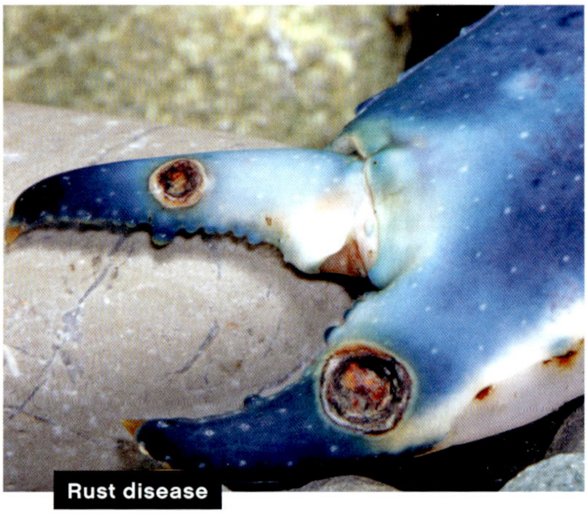
Rust disease

tial symptoms are not unlike the beginning of a moult. Soon afterwards, though, the crayfish will start to show paralyses, stagger instead of walking, become listless and may even lose limbs. After only a few days, it dies. If you notice a crayfish with these symptoms, please do not wait until it is dead, as the water mould will start searching for a new host once this happens. The pathogen can only survive without a host for a few days and thus releases huge amounts of zoospores. Due to the high number of spores the risk of infection rises to incredible heights in an aquarium and even in ponds, and the probability of an epidemic is very high. If the zoospores do not find a new host within a few days they will die off. Fortunately, *Aphanomyces astaci* does not form permanent cysts. When an infected crayfish dies from other reasons, the water mould will also start to search for a new host. Even healthy-looking crayfish can carry the pathogen and are infectious. Today this disease has found a wide distribution, mainly due to crayfish production in aquaculture and to North-American crayfish being released into natural waters. American crayfish may die of the crayfish plague, too, if their immune system is

Poor water quality may lead to fungal infections of the eggs

overtaxed with other pathogens. In Signal Crayfish, this phenomenon has already been proven in scientific studies.

However, not only live crayfish pass the pathogen on to others, it is also known to spread with other aquatic organisms like snails, mussels, fish and even aquatic plants and the water itself. The spores have even been proven to find their way into other waters clinging to damp nets, containers and tubes. Thus you need to be utterly careful during tank maintenance in order not to endanger your crayfish. They must never be released in nature, of course, and other inhabitants of crayfish aquaria need a quarantine of three to four weeks before they go into another crayfish tank. Crayfish should never be kept in a garden pond either. There is no cure for this disease besides prevention!

Another very dangerous disease is rust spot or shell rot. Here the cray's exoskeleton rots away, which can get critically dangerous if larger parts of the skin are eroded. It can be recognised quite easily as the infected spots on the shell turn black and necrotic. The disease can be controlled or prevented by the presence of brown oak leaves or alder cones in the aquarium. The humic acids contained therein help keep the pathogens in check.

High silicate readings in tap water are often the reason for fungal infections of crayfish eggs or for very low hatch rates. In the trade, several products are available to resolve this problem. For many other crayfish diseases, unfortunately there is no safe diagnosis method yet, and thus hardly any therapy is possible.

## Mating and reproduction

Many North-American crayfish make physical and chemical contact right before mating. The high aggression level of some males and the relative passivity of the females are critical factors when the crayfish first meet. During this process, the females are observed to reject overly aggressive males. Further contact stimulates passive crayfish sexually and helps reduce the aggressive behaviour of the potential partners. Only then, mating is made possible. At first the male gets a hold of the female with his claws and tries to turn her on her back or her side. Then he wraps her body with his walking legs, trying to keep her in that position. If he succeeds, the ventral sides of both partners touch, and the actual mating process can take place. Sperm are transported from the spermatic duct into one of the first two gonopods. The second gonopod works like a piston, pumping the sperm forward. The first gonopod is introduced into the female's annulus ventralis, a small receptacle on the underside of the female. The second gonopod creates sufficient pressure so the sperm are deposited there.

In some species, the female is able to store the sperm for a very long time, and the eggs are laid much later. In Australian crayfish, the mating process works somewhat differently. Some days before mating, the female is busy preparing. In the course of this process she uses her last two walking leg pairs to clean her pleopods and the underside of her carapace between the walking legs. A few hours before the actual mating process, the male starts losing his aggressive behaviour, and he starts stroking and touching the female tenderly with his antennae. Shortly before mating, both partners display this tender ritual with their antennae, and the female basically lies down on her back all by herself, letting the male climb on her. The female then folds her pleon under her carapace, and the male covers it with his abdomen. The male deposits the sperm on the underside of the female abdomen or between the last walking leg pair. After that, the pair separates, and the male goes away. Then the female presses out the blueish eggs, inserting them into a gelatinous package.

## Parastacidae family

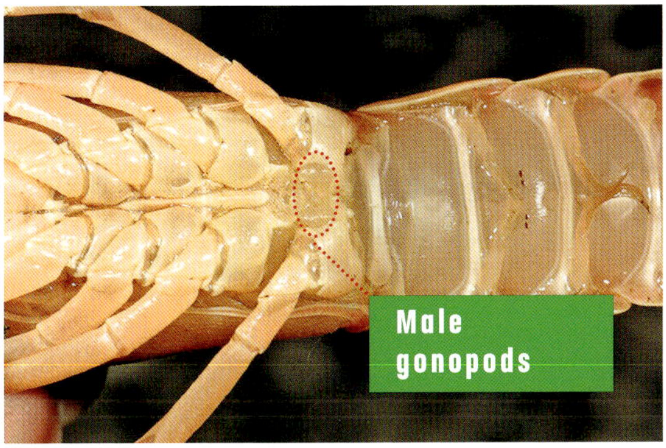

## Astacidae family

## Cambaridae family

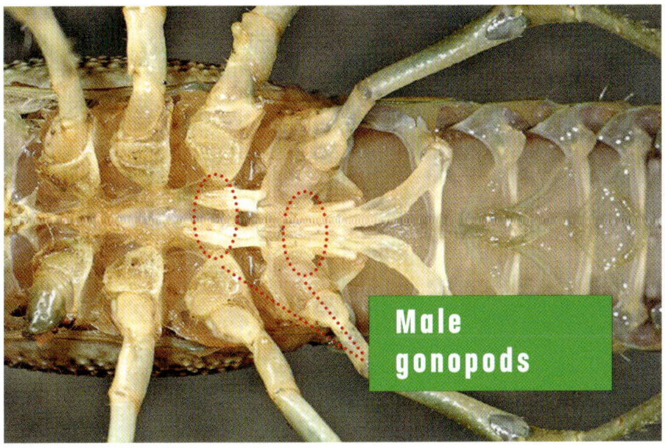


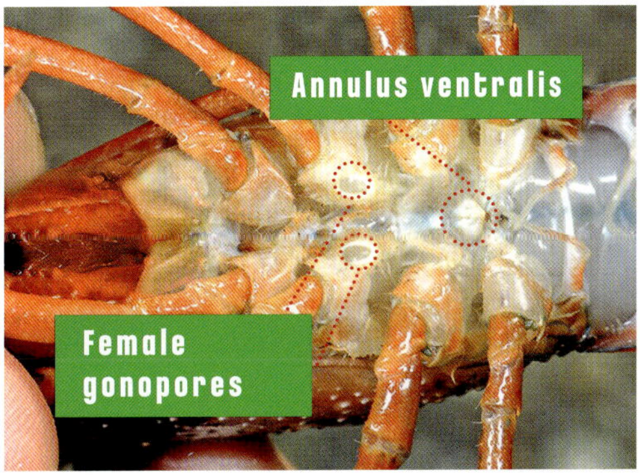

## CRAYFISH

The maximum body length of a crayfish is measured from the tip of the rostrum to the end of the tail fan. The claws are left out as they are of a different size in most crayfish, may have been replaced one or even several times, or vary in form depending on the season of the year. Besides the total body length, sometimes you will also find the term "carapace length". Here only the back of the cephalothorax is measured, from the tip of the rostrum to the dent at the end of the carapace.

Cold River in Missouri (USA)

## NORTH-AMERICAN AND MEXICAN CRAYFISH

The species available in the trade usually belong to the genera *Cambarus*, *Orconectes* and *Procambarus*. Most of them have interesting colours and beautiful patterns. The overwhelming majority of representatives of the genus *Procambarus* are extremely robust and are known to adapt well to a multitude of environments. Most of the species of the genera *Orconectes* and *Cambarus* originate from rivers or creeks in temperate climate zones. Those species need cooler water during the winter months for their well-being and for successful reproduction. There are species of the genus *Cambarus* that ought to be kept by more experienced hobbyists and breeders as they are aggressive and prefer to live their life in hiding. They are not suitable for a traditional social aquarium. These crayfish are avid burrowers. In the wild, they usually hide under rocks or even dig small tunnels into steep river banks where they take cover. North-American an Mexican crayfish are illegal in some countries.

## *Cambarus manningi*

Greensaddle Crayfish

These crayfish reach a total length of five to six centimetres (2 to 2.3"). They are thus suitable for smaller tanks, too. If you want them to reproduce, make sure they can go into hibernation during the winter months. The tank must not get too cold, though. Avoid freezing at all cost. A cool phase at temperatures of 7 to 8 °C (45 to 46 °F) over the course of some weeks is sufficient. In their natural habitat, these crayfish live almost exclusively in rapidly flowing sections of creeks and smaller rivers.

Egg batch

Larva

Body size: 5 to 6 cm (2 to 2.3")
Temperature: 6 to 22 °C (43 to 72 °F)
pH value: 6 to 8
Tank size: 50 cm (20") and more
Distribution: Alabama, Georgia
Reproduction: cold phase required

## *Cambarus rusticiformis*

Depression crayfish

*Cambarus rusticiformis* stays quite small for a crayfish of the genus *Cambarus*. It has attractive colours and patterns and is probably one of the most beautiful North-American crayfish.

In its area of distribution it can be found in small rapidly flowing creeks, where it usually hides under rocks. The adult size is only eight to nine centimetres (3 to 3.5"). These crayfish may vary in colour considerably. Usually their carapace is of a yellowish white, as is the pleon, which has black horizontal bands in addition. The telson is of an intense blue, as are the legs. However, there is also a yellow-green-black variant, which is much more rarely found in the trade.

Body size: 8 to 9 cm (3 to 3.5")
Temperature: 10 to 25 °C (50 to 77 °F)
pH value: 4.5 to 8.5
Tank size: 100 cm (40") and more
Distribution: south-east of the US
Reproduction: cold phase required

## *Orconectes durelli*
### Saddle Crayfish

The upper side of this crayfish is yellow to dark brown, with two lateral bands on its carapace, one cream-coloured and one black band. Especially striking characteristics are the bright orange to red claw tips. *Orconectes durelli* grows to a total length of seven to eight centimetres and prefers clean waters with a rapid current. A flow pump is thus highly recommendable in a tank holding these beautiful crays. If you want them to reproduce make sure they can go through a period of hibernation for some weeks during which they ought to be kept at temperatures of 8 to 10 °C (46 to 50 °F).

Body size: 7 to 8 cm (2.8 to 3.1")
Temperature: 4 to 22 °C (39 to 72 °F)
pH value: 6 to 8
Tank size: 60 cm (24") and more
Distribution: Kentucky, Tennessee
Reproduction: cold phase required

## *Orconectes limosus*
### Eastern Crayfish, Delcore Crayfish

Male and female Eastern Crayfish grow to about the same size, and there are no obvious bodily differences between the sexes either. The males do not even have stronger claws than the females. Originally, this species was distributed throughout the east of North America, however, it was introduced into Europe as early as 1890 in order to make up for economical losses due to the ravaging crayfish plague and in order to replace the native European crayfish that were feared to be dying out. Unfortunately, this has proven to be a critical error, as in our days, North-American crayfish are one of the most dangerous threats to native European species.

Mating takes place in autumn, when water temperatures drop. During copulation, the crays turn their abdomens towards each other. The males cling to the skin on the females' legs with the hooks on their ischia. During the mating process, the crayfish

often lie listlessly for a long time. We were able to observe mating durations of up to 24 hours. After the copulation, no spermatophores are visible on the female. She presses out the eggs many months later (around April). Then the eggs develop rather quickly as water temperatures rise, and the young crays hatch around May (depending on the water temperature). Aggressive behaviour and jumpiness in Eastern Crayfish may be due to a lack of protein in their food. Other than that, they have no special requirements regarding their environment. They even survive in heavily polluted European rivers, after all.

In no case should they be socialised with other crayfish, not even with larger species like the Signal Crayfish. The smaller Eastern Crayfish will kill even large specimens during their next moult at the latest.

Body size: 10 cm (4")
Temperature: 4 to 26 °C (39 to 79 °F)
pH value: 7.5 to 8.5
Tank size: 80 cm (30") and more
Distribution: east of North-America and Central Europe
Reproduction: cold phase required

## *Procambarus alleni*

Blue Florida Crayfish, Everglades Crayfish

One of the most widely spread crayfish in the hobby is the Blue Florida Crayfish, often falsely called Blue Lobster. It is a colour variant of a normally rather unspectacular, brown species. The blue colour is due to an enzyme defect. The males have somewhat narrower claws and grow a bit larger than the females. The appearance of the females is somewhat bulkier.

Smaller females carry, as a rule, 100 to 150 eggs, larger females will have up to 350 young crays per batch. The young crays, in turn, can reach sexual maturity with a body length of only six to seven centimetres. After hatching, they grow relatively fast at first, and moult every few days. In their development, not only food plays a crucial role, but also water temperatures, which ought to be within a range of 20 and 27 °C (68 to 81 °F). As this species is quite cannibalistic the offspring will eat each other, especially freshly moulted young crayfish are predated upon. Thus you'll need a tank with as big a footprint as possible in order to raise the offspring successfully, with lots of places to take cover. Plants will sooner or later be eaten by the crays, and their remainders will clog the filter intake. The only plants reported to be less tasty for crayfish are *Cryptocoryne usteriana* and Java Fern (*Microsorum pteropus*).

Body size: 8 to 10 cm (3.1 to 4")
Temperature: 18 to 30 °C (64 to 86 °F)
pH value: 7.5 to 8
Tank size: 80 cm (30") and more
Distribution: southern Florida
Reproduction: no cold phase required

Procambarus alleni, wild type

In drainage ditches like these in Florida, Procambarus alleni are found

*Procambarus clarkii* habitat in Mississippi

## *Procambarus clarkii*

Red Swamp Crayfish

The original habitat of *Procambarus clarkii* was from northern Mexico to Florida in the east and Illinois in the north. Meanwhile, however, the species has spread to over 50 countries in all the world, as it is far more adaptable than most other crayfish and can live in the most different biotopes. In Europe, many pet shops have been selling *Procambarus clarkii* as crayfish for aquaria and ponds. In the following years, undesired populations established themselves in nature as some crayfish were inadvertently released into the wild. In their areas of distribution, these crayfish can be observed to leave the water after heavy rainfall and to walk over land for long distances. On moist nights, they may cover a length of up to 200 metres (220 yards). They are also known to be true escape artists from aquaria. As a rule, *Procambarus clarkii* live to two or three years of age in the swamps in the south-west of the USA. In captivity they can grow somewhat older if cared for properly, to three or four years of age. Although adult crayfish can reach a body length of up to 15 centimetres (6"), most will stay a bit smaller. In a temperate tank, they have offspring all year round. The time the females stay berried depends very much on the water temperature, too. Usually, the eggs need two to three weeks to develop. The females carry eggs two to three times per year, and each time, 50 to 300 young crayfish will hatch from one batch. The young crayfish grow fast. At first, they are grey to slightly reddish in colour.

*Procambarus clarkii* is said to be especially aggressive and ought to be kept in tanks with a length of one metre or longer. Cohabitation with fish or other crayfish species may cause severe problems and is thus not recommendable.

Body size: 14 cm (5.5")
Temperature: 4 to 36 °C (39 to 97 °F)
pH value: 6 to 8.5
Tank size: 100 cm (40") and more
Distribution: distributed in all the world
Reproduction: unproblematic,
no cold phase required

# Colour morphs of *Procambarus clarkii*

*Procambarus clarkii* "White"

*Procambarus clarkii* "Blue"

*Procambarus clarkii* "Orange"

*Procambarus clarkii* "Ghost"

*Procambarus clarkii* "Tricolour"

*Procambarus clarkii* "Wild type/Texas"

## *Procambarus enoplosternum*

Black Mottled Crayfish

These crayfish mainly live in shallow waters with a slow current. They lead secretive lives as they are well-camouflaged in the waters of these rivers, which are usually of a very dark colour due to lots of fallen leaves. In the habitat of the Black Mottled Crayfish, the water is slightly acidic and on the cool side, however, in captivity they can be kept at somewhat different parameters, too. They are more peaceful than most *Procambarus species*, however, you ought to have a tank with a length of at least 80 cm (31") for a breeding pair.

Body size: 5 to 6 cm (2 to 2.4")
Temperature: 10 to 24 °C (50 to 75 °F)
pH value: 6 to 7
Tank size: 60 cm (24") and more
Distribution: Alabama, Georgia
Reproduction: no explicit cold phase required, ought to be kept somewhat cooler during wintertime, though

## *Procambarus fallax f. virginalis*

Marbled Crayfish, Marmorkrebs

The Marbled Crayfish is not as spectacular in colour as other crays, however, it has a nice marbled pattern whose colouring depends on the water parameters. In calcium-rich water they are of a greenish brown, in acidic water they assume a blueish hue. Their most outstanding characteristic is their ability to reproduce without males, which is unique in the freshwater decapod world. Only females are available in trade, and they reproduce parthenogenetically, which means that new life is generated from an unfertilised egg cell. Thus, this crayfish has absolutely no problems reproducing, and as it is a highly adaptable species, any aquarium may soon become too small. It is best to use the offspring as food for other animals like turtles or large cichlids. The Marbled Crayfish is generally quite peaceful and does not predate on other aquarium inhabitants. From time to time it can develop cannibalistic traits and eat its own offspring or other conspecifics.

This species can reach a body length of 8 to 12 cm (3 to 4.7"). Its temperature range is large, but it is best kept at room temperature. As the Marbled

Crayfish may survive in nature in temperate climate zones and as several independent populations have been observed in nature in Central Europe, crayfish of this species must not be released into the wild under any circumstances!

Body size: 8 to 12 cm (3.1 to 4.7")
Temperature: 4 to 36 °C (39 to 97 °F)
pH value: 6 to 8.5
Tank size: 60 cm (24") and more
Distribution: worldwide in aquatics, released into nature in Germany and Madagascar
Reproduction: no cold phase required

## *Procambarus spiculifer*

### White Tubercled Crayfish

One of the larger representatives of the genus, *Procambarus spiculifer*, prefers cool, slow-flowing black water and fine sandy ground. In nature, these crayfish are mainly found in places covered in wood or leaves, sometimes also under rocks. In captivity, their aquarium ought to be decorated with similar materials.

Their dark brown claws are covered with striking ivory-coloured tubercles. *Procambarus spiculifer* is easily mistaken for similar species like its close relative *P. gibbus* or *P. echinatus*. A breeding pair needs a tank with a length of 100 centimetres (39") or more. The White Tubercled Crayfish can be kept at room temperature without any problems, and it tolerates cooler temperatures in winter as well as high temperatures during the warm season.

These crayfish can be kept together with robust dwarf shrimp and sucker-mouth plecos. Other fish may be in some danger during the night when they sleep and the crays go searching for food.

Body size: 12 to 14 cm (4.7 to 5.5")
Temperature: 10 to 25 °C (50 to 77 °F)
pH value: 4.5 to 8.5
Tank size: 100 cm (40") and more
Distribution: south-east of the US
Reproduction: no explicit cold phase required, ought to be kept somewhat cooler during wintertime, though

## *Procambarus llamasi*

### Yellow-banded Crayfish

These crayfish originate from eastern Mexico, from Veracruz to the border of Guatemala in the south. Their claws are tightly covered in bristles and fine hair, which discerns them from most other Mexican or North-American species. *Procambarus llamasi* inhabits rivers and canals, and it is even found in temporary small lakes or ponds. It grows to a total body length of 8 to 9 centimetres (3 to 3.5"). One sex always prevails in a batch of eggs. Presumably, temperatures during the brooding period of the female play an important role in the outcome, comparable to some other invertebrates. The Yellow-banded Crayfish is relatively peaceful and can be kept in small groups. Fish, plants or dwarf shrimp are basically left alone. The bristle cover on the claws of wild-caught crayfish is much denser than in tank-bred specimens. Sometimes it disappears entirely in crayfish hatched in captivity. Up until now it is still quite unclear which function these bristles fulfil.

Body size: 8 to 9 cm (3 to 3.5")
Temperature: 18 to 28 °C (64 to 82 °F)
pH value: 6 to 8
Tank size: 80 cm (30") and more
Distribution: Veracruz, Mexico
Reproduction: no cold phase required

## *Procambarus vazquezae*
### Veracruz Crayfish

This relatively small representative of the genus will not grow larger than 5 or 6 centimetres on average, which makes *Procambarus vasquezae* a good crayfish for smaller aquaria. Relative to the body, its claws are of a striking size. It is suitable for social

tanks with larger livebearers, characins and catfish. Its behaviour is not very typical for a *Procambarus* but comparable to that of the genus *Cambarellus*, as it is quite peaceful towards conspecifics and can thus be kept in groups. Even adult males have been observed to peacefully sit in a cave together. The females carry eggs three to five times a year, given that they live in good conditions, and as these crayfish are virtually non-cannibalistic the offspring can be kept in the same tank as the parents, which makes this species especially suitable for beginning crayfish enthusiasts. In trade, you can often find blueish-grey or yellowish *P. vasquezae*. Their colour depends very much on the food and the water parameters. Plants are not damaged or eaten.

Body size: 5 to 6 cm (2 to 2.4")
Temperature: 18 to 28 °C (64 to 82 °F)
pH value: 7.5 to 8
Tank size: 60 cm (24") and more
Distribution: Veracruz, Mexico
Reproduction: no cold phase required

# RARE CRAYFISH FROM NORTH AMERICA

*Cambarus englishi*

*Cambarus scotti*

*Cambarus luteus*

Orconectes palmeri longimanus

Orconectes medius

Orconectes palmeri palmeri

CRAYFISH

*Cambarellus montezumae* habitat in Mexico

## North-American and Mexican Dwarf Crayfish

Dwarf crayfish of the genus *Cambarellus* are best kept in a group and in a dedicated tank, as most fish would predate on their offspring. Of course, a higher survival rate means more offspring, which in turn has the consequence that the young crayfish may cannibalise each other. Older conspecifics will also predate on the young ones. To prevent these kinds of losses it is very important to provide the crayfish with a balanced diet rich in protein.

In a dwarf crayfish tank, a heater is not absolutely necessary, as fluctuations in water temperature also occur in their natural habitats, and have a beneficial effect. The claws of the females in almost all species of the genus *Cambarellus* are shorter and thicker than those of the males. Females usually grow older than the males, under good circumstances they can reach an age of two to three years, given that they live in good conditions. They can have up to four batches of eggs per year and will thus have plenty of offspring.

As dwarf crayfish are diurnal invertebrates, you will often see them copulating. The mating process takes around 15 minutes. After the act, the female is released from the tight grip of the male, and the crayfish separate speedily with some quick strokes of their pleons. After around three weeks, the fully developed young crayfish will leave their mother to start life on their own.

## *Cambarellus diminutus*

Least Crayfish

In nature, *Cambarellus diminutus* are found in small stagnant waters, in creek or river zones with a slow current or in roadside ditches. Most frequently you can find them under detritus or in dense vegetation, where they take cover. Increasing human settlement and the destruction of their habitats have caused the Least Crayfish to be critically endangered in the wild. This crayfish species has adapted to higher temperatures of over 30 °C (86 °F). Its carapace length is 1.2 centimetres (0.5"), and its total body length 2.5 to 3 centimetres (1 to 1.2").

We have found this species to be quite variable in colour, from light blue to reddish brown. There is hardly any intraspecific aggression. The females usually carry a batch of 15 to 20 eggs several times a year in the wild as well as in the aquarium.

Body size: 2 to 3 cm (0.8 to 1.2")
Temperature: 18 to 30 °C (64 to 86 °F)
pH value: 6 to 8.5
Tank size: 20 cm (8") and more
Distribution: USA, Alabama and Mississippi
Reproduction: no cold phase required

## *Cambarellus ninae*
### Arkansas Dwarf Crayfish

One of the most peaceful dwarf crayfish species that were introduced into the aquarium hobby during the last years. *Cambarellus ninae* are very active even during the day. They grow to a length of up to 3 centimetres (1.2"), the females being a bit larger in size than the males. Their colour varies from dark over light brown to a whitish or blueish grey. Some specimens are even dark blue.

In their natural habitat, the coastal plains of Texas and the flatlands of Arkansas, the winters can be rather cold, and temperatures range from 5 or 6 °C (41 to 43 °F) during the winter to up to 30 °C (86 °F) during the summer months. Their reproduction rate is quite high, since the mother crayfish carries a new batch of eggs again only a few days after her offspring have hatched. There are approximately 70 to 80 young crayfish per batch. When kept in good conditions, the offspring reach sexual maturity after as short a period as three months.

*Cambarellus ninae* can be socialised with small fish in a planted tank without any problems. Their high reproduction rate requires a somewhat larger tank, though, in order to provide them with sufficient space.

Male

Body size: 2.5 to 3 cm (1 to 1.2")
Temperature: 6 to 30 °C (43 to 86 °F)
pH value: 7.5 to 8.5
Tank size: 30 cm (12") and more
Distribution: Texas, USA
Reproduction: no cold phase required

Female

## *Cambarellus puer*

Swamp Dwarf Crayfish

Female

This species was introduced into the aquarium hobby in 2002. Meanwhile, it has found a great number of friends, and it is successfully bred in captivity. At 3.5 to 4 cm (1.4 to 1.6"), the females grow a bit larger than the males. With around two years, their life expectancy is rather short.

In nature, *Cambarellus puer* is mainly found in densely overgrown muddy zones near the edges of waterbodies, where they take shelter from predators. Berried females can be found almost all year round. On average, they carry 40 to 110 eggs, the exact number of which depends strongly on the size of the female. This crayfish has a large area of distribution, which makes differences in pattern, colour and morphology highly probable. As for all dwarf crayfish, the Swamp Dwarf Crayfish aquarium ought to be well-stocked with oak or beech leaves – an important factor for their well-being and for successful breeding.

Body size: 2 to 4 cm (0.8 to 1.6")
Temperature: 18 to 30 °C (64 to 86 °F)
pH value: 6 to 8.5
Tank size: 20 cm (8") and more
Distribution: Texas to Missouri, USA
Reproduction: no cold phase required

Male

## *Cambarellus patzcuarensis*

Orange Crayfish, CPO

Female

This is the most frequently kept dwarf crayfish in our aquaria, and especially its orange colour morph, in short called CPO (for *Cambarellus patzcuarensis* "orange"), can be found in almost every well-sorted local fish store. In some countries though, these Crayfish are illegal! Its wild form is more of a light to dark brown. Many specimens – of both sexes – have darker stripes on the upper side of their bodies, some have light and dark spots, and some everything in between. In the orange colour morph, these patterns will occur, too.

These crayfish are frequently found in dense water plant populations near the shore of Lake Patzcuaro in Mexico. The females grow a bit larger than the males and also somewhat older. Under good conditions, they can reach two years of age. During this time, they can get berried up to four times, thus producing large numbers of offspring. Depending on the size of the female she will carry 25 to 50 eggs under her pleon. The reason for fungal infections of the eggs is often found in a high silicate reading of the water in the tank. In the trade, there are remedies against this problem.

In almost every batch of CPO eggs, some are unfertilised. They can easily be discerned from good ones by their yellowish-orange colour, whereas viable eggs are dark brown. Usually, the females will remove unfertilised eggs in order to prevent the entire batch of eggs from developing a fungal infection.

Brown colour morph

Body size: 2 to 4 cm (0.8 to 1.6")
Temperature: 15 to 30 °C (59 to 86 °F)
pH value: 7.5 to 8.9
Tank size: 30 cm (12") and more
Distribution: Mexico, Lake Patzcuaro
Reproduction: no cold phase required

Female with larvae

Berried female

Male

## *Cambarellus schmitti*

Fontal Dwarf Crayfish

This relatively small *Cambarellus* species prefers clear flowing waters, and only rarely can it be found on muddy or sandy substates. Usually they take cover in detritus. Even though these crayfish originate from Florida they ought not be kept too warm. At room temperature, a heater is not required. They only reach 1.5 years of age, which is indeed not much, however, during this time, a female can have up to three batches of offspring. In their natural habitat, berried females can be found all year round. *Cambarellus schmitti* is a relatively peaceful crayfish, and can live in groups. Disputes are few and far between, and usually do not have any grave consequences. Fish and plants are left alone, however, these crayfish should only be kept together with small fish in order to prevent the tiny offspring from becoming fish food.

Habitat in Florida (USA)

Body size: 2.5 to 3 cm (1 to 1.2")
Temperature: 18 to 23 °C (64 to 73 °F)
pH value: 7.5 to 8
Tank size: 30 cm (12") and more
Distribution: Florida, USA
Reproduction: no cold phase required

## *Cambarellus texanus*

Brazos Dwarf Crayfish

This species mainly inhabits smaller ponds in the state of Texas, and even some highly polluted roadside ditches hold extremely dense populations of the Brazos Dwarf Crayfish. Temperatures may vary from 4 °C (39 °F) in winter to over 30 °C (86 °F) in summer. When stagnant habitats of *C. texanus* fall dry, which is quite often the case during the dry months in Texas, the crayfish dig small burrows into the soft, overgrown ground in order to flee the drought. These crayfish are of a uniform brown or light brown colour. Other than in other *Cambarellus* species, we have not been able to find any striped specimens of *C. texanus*.

In the aquarium, these crayfish are rather peaceful and seem to do well even though water parameters may vary. They are also very uncomplicated to breed. Females carry 40 to 50 eggs. After two to three weeks, they release their offspring, which start living independently in the aquarium at once.

Habitat in Texas (USA)

Body size: 2 to 3 cm (0.8 to 1.2")
Temperature: 4 to 30 °C (39 to 86 °F)
pH value: 7.5 to 8.5
Tank size: 30 cm (12") and more
Distribution: Texas, USA
Reproduction: no cold phase required

## *Cambarellus shufeldtii*
Cajun Dwarf Crayfish

Local variant from the USA

*Cambarellus shufeldtii* is a small reddish-brown to grey crayfish with dark longitudinal stripes or dots arranged in irregular rows. Thus we could say there are two variants, occurring in both sexes: The striped and the marbled or spotted morph.

Striped variant

Their adult size is 2 to 3 centimetres (0.8 to 1.2"), however, in good conditions they can grow to a length of up to 4 centimetres (1.6") in the aquarium. These crayfish are diurnal, inhabit the most different types of fresh water bodies, and prefer sitting in aquatic plants. They do not burrow and leave the aquarium substrate alone. The lifespan of a female is around one year. During this time, it can reproduce twice. The males get a bit older than the females, from 15 to 18 months of age. They reach sexual maturity somewhat later than the females, and in the aquarium they grow even older.

The female carries around 30 to 40 eggs, which need approximately three weeks for development. During this time, they are attached to the female's abdomen where they are looked after diligently by their mother. As the larvae and young crayfish are very small, they are easy prey for almost all fish. Small tanks starting at 30 litres of volume (8 US gallons) are suitable for a breeding group. A pair can be kept in a 12 litre tank (3 US gallons) for a short time. Their natural area of distribution is very large, and they live in different water parameters in nature. In their natural habitats, temperatures may

Habitat in Illinois (USA)

vary from 1 to 30 °C (34 to 86 °F), an average of 18 to 22 °C (64 to 72 °F) is recommended in the aquarium. If the crayfish are relatively young at the time

Colour variant

you buy them they will adapt more easily to different water parameters and generally cope better with the stress of moving than fully grown adults. Their small size and their peacefulness make them optimally suited for aquaria, as they leave plants and fish completely alone. All sorts of fish food offered by the trade can be used for feeding them. Snails are a welcome addition to their diet. They can reduce a snail infestation in an aquarium within a short time and will keep the population in check. The tiny snails of the genus *Physa* that annoy so many aquarium keepers are preferred food for these crayfish.

Body size: 2 to 4 cm (0.8 to 1.6")
Temperature: 4 to 30 °C (39 to 86 °F)
pH value: 6 to 8.5
Tank sizc: 30 cm (12") and more
Distribution: Texas, USA
Reproduction: no cold phase required

## Crayfish from New Guinea

It is located south of the Equator and only 160 kilometres to the north of the Australian Cape York Peninsula: the second largest island in the world, New Guinea, a truly exotic paradise full of contrasting landscapes, and one of the last true wildernesses in the world. Vast parts of the island are still unexplored, and thus scientists find new species in the region quite frequently.

Most crayfish from New Guinea were identified and named by the scientist L. B. Holthuis from the natural history museum in Leiden. In 1949, Holthuis was able to trade beads or razor blades for colourful crayfish in order to come into possession of several undescribed species, which were brought to him by local fishermen.

## *Cherax lorentzi*

Black Tiger Crayfish, Lorentz' Crayfish

Like male Red Claw Crayfish (*Cherax quadricarinatus*), male *Cherax lorentzi* have a bright red patch on the outer margin of their claws. *Cherax lorentzi* have stronger claws in comparison. Females do not have a red patch at all. They are known to breed successfully in captivity, however irregularly. Berried females retreat to a hide-out and can hardly ever be seen. This crayfish reaches sexual maturity at a body length of only five to six centimetres (2 to 2.4"). If you do not separate the parents from their offspring after the latter have become independent, only a few will survive. Temperatures of 20 to 24 °C (68 to 75 °F) are considered the optimum range, temperatures of under 15 °C (59 °F) are tolerated for only a very short time. They should not be kept in too soft water. *Cherax lorentzi* is unaggressive towards other tank inhabitants like fish or shrimp, and it does not even harm bottom-dwelling catfish. Aquatic snails are considered food. Plants may be damaged as the crayfish dig, however they do not eat them to an extent that robust, fast-growing plants would not survive in the tank.

Body size: 7 to 12 cm (2.8 to 4.7")
Temperature: 20 to 24 °C (68 to 75 °F)
pH value: 7.5 to 8.5
Tank size: 60 cm (24") and more
Distribution: Western Papua
Reproduction: no cold phase required

## *Cherax boesemani*
Red Brick Crayfish,
Blue Brick Crayfish

Their body is often of a brick red to reddish brown, and the claws of adult *Cherax boesemani* are blueish black and brick red and end in black tips. On the inner margin, the claws have pink or reddish lines that look like a serrated band. There

are also specimens with blue claws, a greenish body and a reddish pleon. On the outward margin of the claws, the males have whitish patches, the females don't. These crayfish can grow to an impressive size, however, in the aquarium they are rather peaceful and calm, also towards conspecifics. They can be kept at room temperature in an unheated tank all year round, as water temperatures in their original habitat are also rather on the cooler side, at only 19 to 21 °C (66 to 70 °F). *Cherax boesemani* can be socialised with small fish like livebearers as well as with dwarf shrimp. *Ancistrus* catfish can also live in a tank with these crayfish very well. They hardly eat aquatic plants at all, however, root plants can be damaged as they burrow actively.

Body size: 20 to 25 cm (7.8 to 9.8")
Temperature: 19 to 21 °C (66 to 70 °F)
pH value: 7.5 to 8
Tank size: 100 cm (39") and more
Distribution: Western Papua
Reproduction: no cold phase required

CRAYFISH

## *Cherax pulcher*

Hoa Creek Crayfish, Irian Jaya Crayfish

The Hoa Creek Crayfish is one of the most colourful *Cherax* species. The inner side of its claws is blue, their outward margin is white, and the males have an additional white patch. In general, the males have more intensive colours than the females. There are several colour variants of this species.

Unfortunately, not much is known about their life in their natural habitat. The Hoa Creek Crayfish is peaceful towards conspecifics and fish and shrimp in the tank. Adult males may run into one or other disagreements with each other, though. They create many burrows, thus plants will have difficulties in a Hoa Creek Crayfish tank and are best attached to driftwood or rocks.

Body size: 9 to 12 cm (3.5 to 4.7")
Temperature: 20 to 24 °C (68 to 75 °F)
pH value: 6.5 to 7.5
Tank size: 80 cm (30") and more
Distribution: Bird's Head Peninsula, Western Papua
Reproduction: no cold phase required

## *Cherax* sp. "Blue Moon"

Blue Moon Crayfish

Both sexes of the Blue Moon Crayfish are of a beautiful dark blue to almost black from head to telson. There is a stunning thin white stripe across each segment of their pleon. Adult males have a white patch on the outward margin of their claws (comparable to the red patch *C. quadricarinatus* or *C. lorentzi* have). The margin of the tail fan is of a striking orange colour, and the carapace has some light blue to white dots on its sides.

The Blue Moon Crayfish is rather peaceful with conspecifics and other tank inhabitants, with the exception of adult males that will fight each other for territorial reasons. Given that there is enough room the males will cease fighting as soon as the boundaries of their territories are clear. These crayfish do not tolerate temperatures below 15 °C (59 °F). They burrow a lot, thus plants will have difficulties in a Blue Moon Crayfish tank.

Body size: 10 to 13 cm (3.9 to 5.1")
Temperature: 20 to 25 °C (68 to 77 °F)
pH value: 7.5 to 8
Tank size: 100 cm (40") and more
Distribution: Bird's Head Peninsula, Western Papua
Reproduction: no cold phase required

## *Cherax holthuisi*
### New Guinea Apricot Crayfish

*Cherax holthuisi* is very variable in colour, from intense orange to greyish, light yellow and even whitish, and there are even greenish-orange specimens with striking orange claw tips. In the aquarium, these crayfish are quite shy and live a secluded life. They can only be seen searching for food in the evening or during the night. They are avid burrowers and hide under rocks or driftwood. They hardly eat plants, given that you only use robust variants in the tank. If they do not find a sufficient number of places to hide they may dig under plant roots to make a tunnel. This species does not tolerate temperatures below 20 °C (68 °F). They reproduce regularly in captivity even though the females are berried for a relatively long time. Even at rather high temperature levels of 24 to 25 °C (75 to 77 °F) the young crayfish hatch after seven to nine weeks. They grow slowly after hatching, quite similarly to the offspring of crayfish from cold water habitats, even though they live at much higher temperatures all year round. After as long as one year they have reached a body length of 3 to 5 centimetres (1.2 to 1.9"). Even though *Cherax holthuisi* have much more offspring per batch than other *Cherax* species, their population grows slower in comparison as there are longer periods of time between batches.

They usually do not predate on fish, and dwarf shrimp as well as ground-dwelling fish have nothing to fear from these crayfish as a rule either. Should you notice them going after fish nonetheless, make sure you increase the amount of protein you feed.

Body size: 9 to 12 cm (3.5 to 4.7")
Temperature: 20 to 24 °C (68 to 75 °F)
pH value: 6.5 to 7.5
Tank size: 80 cm (30") and more
Distribution: Bird's Head Peninsula, Western Papua
Reproduction: no cold phase required

CRAYFISH

Fishing for crayfish on Lake Paniai

# CHERAX HOLTHUISI COLOUR MORPHS

"Blue" variant

"Ice blue" variant

"Chocolate-coloured" variant

"Green Orange Tip" variant

"Orange Tip" variant

"Light brown" variant

## *Cherax* cf. *holthuisi* "Black Irianto"

Black Scorpion Crayfish,
Black Irianto Crayfish

The first Black Scorpion Crayfish were imported in 2007, however, it is not quite clear yet which species they belong to or whether their species is still undescribed. Their morphology is quite similar to that of *Cherax holthuisi*, however, the eyes are much larger in *Cherax* cf. *holthuisi* "Black Irianto".

In the aquarium, they are more active and more aggressive than *Cherax holthuisi*, however, they do share other behavioural characteristics. They have been bred in captivity. Depending on their body size, the females carry 40 to 80 eggs, with a relatively long development time of nine to ten weeks at room temperature. The young crayfish grow very slowly, some of them do not even grow to a body length of 5 centimetres (1.9") in a year.

Body size: 9 to 12 cm (3.5 to 4.7")
Temperature: 20 to 24 °C (68 to 75 °F)
pH value: 6.5 to 7.5
Tank size: 80 cm (30") and more
Distribution: Bird's Head Peninsula, Western Papua
Reproduction: no cold phase required

During our travels, like here on Papua, members of the populace often give us important and valuable information about the crustaceans in their immediate surroundings. The region around the Wissel Lakes in central Western Papua was closed off to foreigners for years due to the long-lasting civil unrest in the area. However, in 2010 we were finally able to go on a scientific expedition in this remote corner of the world in order to find out more about the inverts of the region.

## *Cherax peknyi*

### Zebra Crayfish, Tiger Crayfish

Zebra Crayfish

The Zebra or Tiger Crayfish is quite shy when kept in an aquarium, and it leads a very secretive life. Usually, it leaves its hiding-place only at night or when fed special treats. It is a dedicated burrower, and it frequently re-decorates its tank radically when building new tunnels. A thick layer of sand or gravel, stabilised by driftwood and rocks, will provide this species with the ideal conditions to build its own burrows.

Despite their impressive length of 10 to 12 centimetres (3.9 to 4.7"), these crayfish are quite peaceful. They do not predate on fish, given that their food contains sufficient amounts of protein. *Cherax peknyi* seem to love snails, though, and will eat the entire snail population of a tank within a very short time. If the tank is overstocked, large males may become aggressive towards other males, whose limbs may be compromised.

For breeding, a calm tank with a small, if any, stock of peaceful fish is ideal. Berried females retreat into caves. During the time they carry eggs, which may take four to five weeks at a temperature of 23 to 25 °C (73 to 77 °F), the female hardly ever leaves her hiding-place. The offspring of *Cherax peknyi* grow very slowly and are also very secretive. These crayfish hardly ever eat aquatic plants.

Body size: 10 to 12 cm (3.9 to 4.7")
Temperature: 20 to 24 °C (68 to 75 °F)
pH value: 6.5 to 7.5
Tank size: 80 cm (30") and more
Distribution: Papua-New Guinea, Western Papua
Reproduction: no cold phase required

Blue Tiger Crayfish

Red Tiger Crayfish

These crayfish are sold as Zebra or Tiger Crayfish in the trade. The Tiger variant has significantly wider stripes on the upper side of its pleon than the Zebra variant. The Tiger variant has yellow, the Zebra variant white stripes.

*Cherax* sp. "New"

*Cherax communis*

*Cherax communis*

Cherax vogelkop

Cherax sp.

Cherax monticola

# CRAYFISH FROM AUSTRALIA

Australia – the largest "island" in the world. A country with hot deserts, tropic rain forests and kilometre after kilometre of beautiful lonesome beaches, with an incredible biodiversity of flora and fauna. A country, unreal, beautiful and bizarre, where curious hopping animals raise their young while carrying them in a bag, with birds that can dance, laugh and outrun a car, with beaked animals that live under water yet suckle their offspring, swans that aren't white but black, and with many curious-looking nocturnal animals with large eyes, large ears, and soft fur. All this fascinates the large numbers of tourists that travel to Australia year after year to enjoy its plentiful exotic flair, however, a crayfish enthusiast knows that, in addition, over 130 different Australian crayfish species have been described and that there is still a large number yet to be discovered. The crayfish found in Australia all belong to the Parastacidae family. Besides Australia, members of this family also occur on Madagascar, New Guinea, New Zealand and in South America.

Crayfish habitat in Australia

CRAYFISH

## *Cherax destructor*
Yabby

The Yabby is a crayfish with a highly variable coloration. It can be black, brown, orange, blue, ochre, greenish-blue, olive green or grey-white. Its claws are wide and roundish, with a smooth surface and a lighter-coloured pattern on their upper side. Females tend to stay a bit smaller than males of the same age, which can grow to an impressive size and may weigh up to 300 grams (10 oz). Usually, Yabbies live in turbid waters with an extremely slow current and a very dense vegetation.

In nature, *Cherax destructor* lives for up to four or five years, in captivity they can get much older given that the conditions are good. Females produce around 500 to 1 000 young crayfish per year, depending on their body size. In nature, only a small percentage survive the first year, and even in the aquarium it is hardly possible to raise all of them. After mating, the females go into hiding and live secretively in a cave, which they only leave when it is dark in order to eat some food. Their eggs have an irregular ovoid shape. The mother crayfish cares for them with great diligence. She is berried for about three weeks, depending on the temperature.

The Yabby tolerates a vast range of different temperatures, however, you ought to make sure the water temperature never goes below 16 °C (61 °F) as they would otherwise stop eating and also stop growing. Even though these crayfish have "destructor" in their scientific name they are very peaceful towards conspecifics and other tank inhabitants. They love eating plants, though, and it is very difficult to maintain a planted tank with Yabbies in it.

*Cherax destructor albidus*

Body size: 20 to 25 cm (7.9 to 9.8")
Temperature: 20 to 28 °C (68 to 82 °F)
pH value: 7.5 to 9
Tank size: 100 cm (40") and more
Distribution: Australia
Reproduction: no cold phase required

Cherax destructor "wild form"

Cherax destructor destructor

## *Cherax quadricarinatus*

### Australian Red Claw Crayfish

This crayfish needs a very large tank as it grows exceedingly fast and can reach an impressive size with a body length of over 22 cm (8.7"). Red Claw Crayfish reach sexual maturity after only one year, reproduce within a temperature range of 22 to 32 °C (72 to 90 °F) and carry between 500 to 1 500 eggs per batch – up to three times a year. They can reach an age of six to eight years.

The long red patches on the outward margins of the claws of the males are characteristic. They form when the crayfish has reached a body length of approximately 6 to 10 centimetres (2.4 to 3.9"). These patches are soft as they consist of a thin red or orange-coloured membrane. Scientists think they have a signal effect, telling other crayfish the sex, strength, size and the state of health of its owner.

The Australian Red Claw Crayfish can be socialised very well with small fish and dwarf shrimp. Plants can hardly be kept alive in a tank with *Cherax quadricarinatus*, only floating and free-drifting plant species survive at least a short time.

Body size: 10 to 13 cm (3.9 to 5.1")
Temperature: 20 to 25 °C (68 to 77 °F)
pH value: 7.5 to 8
Tank size: 100 cm (40") and more
Distribution: Northern Australia
Reproduction: no cold phase required

Female with larvae

Larvae

*Cherax cainii* "Blue"

*Cherax cainii*

*Cherax preissii*

# ANOMURANS

Alto Bío Bío bordering on Raico Reserva Nacional, a national park in Chile

# ANOMURANS

Aeglids, all of them members of the genus *Aegla*, belong to the anomurans. Usually, members of this group of invertebrates are found in marine habitats and are kept in marine aquaria in the hobby. The vast majority of the 1700 species hitherto described live in the sea. However, approximately 40 species live in fresh water, and some of them can be bought in aquarium shops. The things we know about the requirements of these fascinating inverts are scarce, and not many reports of successful keepers are available.

Aegla sp. from Rio Ranco (Chile)

All freshwater-dwelling species of the genus *Aegla* are found in temperate and subtropic regions of South America. Only here have they been able to make the evolutionary step from marine conditions to freshwater rivers and lakes. Even though anomurans are found in marine habitats in tropical seas in exceedingly large numbers, none of them are found in tropical freshwater biotopes.

The body of the aeglids resembles partially a crayfish and partially a crab. Despite their anatomy, they do not form an evolutionary link between these orders. Their cephalothorax is not horizontally oval to round like that of a crab, but more longish. Their body is very flat, an adaptation to life under rocks and in crevices. When viewed from above, two to three pleon segments that are not folded under the body strike the viewer as characteristic. This is another difference between aeglids and crabs, whose entire pleon is so small that it is entirely tucked under the cephalothorax and cannot be seen from above at all. Like crabs, however, the aeglids are unable to straighten their pleon. When they feel threatened they use their pleon like a paddle and flee backwards, performing quick strokes towards the front and thus creating a series of repulsions which bring them out of the danger zone.

Like all decapod crustaceans, female *Aegla* carry their eggs under their pleon. The entire development takes place inside the egg, there are no free-floating larval stages needing brackish or marine water for growing up. This has enabled

the aeglids to conquer freshwater habitats far away from the coast. The time the eggs need for development depends on the species and the location where it originates from. In temperate climate zones, aeglids adapt their reproductive cycle to the seasons in order to give their offspring the best possible start when they hatch during the warm months, as they find more food and will have a higher survival rate then.

In contrast to crayfish and freshwater crabs, which have four walking leg pairs behind their large chelipeds, *Aegla* only have three walking leg pairs. Of course, the fifth leg pair is there, however, it is much smaller than the others and can thus hardly be seen if you do not especially look for it. Usually it is hidden under the tucked-in pleon. It is equipped with a small pincer and strong bristles. Its function is to clean the inside of the pleon, and the female uses it to examine and clean its eggs when berried. *Aegla* are the only freshwater decapods that carry a pincer on their fifth walking leg pair. The aeglids inhabit different habitats, from vast lakes to rivers and brooks. They live close to the shore or riverbank in shallow water zones, however, they can also be found at depths of over one hundred metres in profound lakes. They live in mountain streams with a strong current as well as in the calm bays of stagnant waterbodies. During the day, they usually hide away under rocks or driftwood, however, they can also be found in algal growth and dense aquatic plant populations. Some are found in creeks with no higher plant growth at all, though.

From time to time, *Aegla* leave the water. Then they sit under driftwood or flat rocks near the water edge in sufficiently moist places. During mating season you will literally find heaps of female aeglids outside the water in these hiding-places. In the aquarium, they will soon cease to be nocturnal only and will also be seen during the day. Maybe this change in behaviour is due to the lack of predators in the tank.

## Keeping *Aegla* in the aquarium

Aeglids are omnivorous and easy to feed in captivity as they accept flake, tablet, frozen as well as live food. Tubifex or bloodworms, which sink to the ground, are among their favourite foods. However, they will also eat snails, and snails of an impressive size, too, that may weigh more than the *Aegla* itself. Aquatic plants are completely or almost entirely left alone, and aeglids can be socialised with smaller fish without any problems, too. In captivity, they are active during the day and get used to their human keeper to the extent that they do not flee from the hand that feeds them, given that there are no large fish that could possibly predate on them. Besides brown *Aegla*, there are some species with intensely orange specimens. The largest representatives grow to a length of five to six centimetres (2 to 2.4"), however, most aeglids do not grow larger than three centimetres (1.2"). In an aquarium dedicated to *Aegla* you'll need to provide lots of hiding-places in the form of rock slabs and driftwood with flat, narrow crevices where the crustaceans can take cover. You'll need to make your aquarium absolutely aeglid-proof to prevent them from escaping, as these crustaceans are known to leave the wet element in nature, too, a behaviour that will usually end fatally in a normal aquarium. It has been proven beneficial to offer them a place where they can sit above the waterline inside the tank, i. e. to keep them in a type of aquaterrarium instead of an aquarium. They only need a small land part.

When keeping *Aegla* in a tank, a crucial point to keep in mind is temperature. All of these crustaceans live in subtropic to temperate climates, and in a common social tank temperatures are too high for them. As creek and river dwellers, they need high oxygen levels in the water, which you need to keep in mind when it comes to water quality. To prevent too fast an increase in nutrients and potentially noxious substances, regular water changes have to be performed.

*Aegla* species are hard to discern from each oth-

Natural *Aegla* habitat in Chile

Cold brook in the Chilean mountains



er, and it is impossible for the layman without any exact knowledge of the location they were collected from. Room temperature is perfect for aeglids from temperate climate zones. Well-structured small tanks with a volume of 30 litres upwards are suitable for keeping a small group of three to four *Aegla*.

A mixture of sand and gravel is best used as substrate. Aeglids do not mind a strong current in their tank, on the contrary, a high water turnover and good filtration are of the essence. When the tank is richly decorated with driftwood and rock slabs you will be able to observe your aeglids climb around. They are also thankful for opportunities to sit above the waterline.

Not much is known about the life expectancy of these crustaceans in captivity. As they can live for several years in nature they are probably long-living aquarium inhabitants that will give their keeper joy for several years, given that they are cared for well. Unfortunately, these rather unproblematic crustaceans have not found many friends in the hobby yet, and they are imported by only a few wholesalers.

# AEGLA

*Aegla platensis*

*Aegla cf. manni*

*Aegla laevis*

*Aegla abtao*

*Aegla pewenchae*

*Aegla occidentalis*

AEGLA

*Aegla araucaniensis*

*Aegla hueicollensis*

*Aegla* sp.

# CRABS

# 21 SPECIES

# CRABS

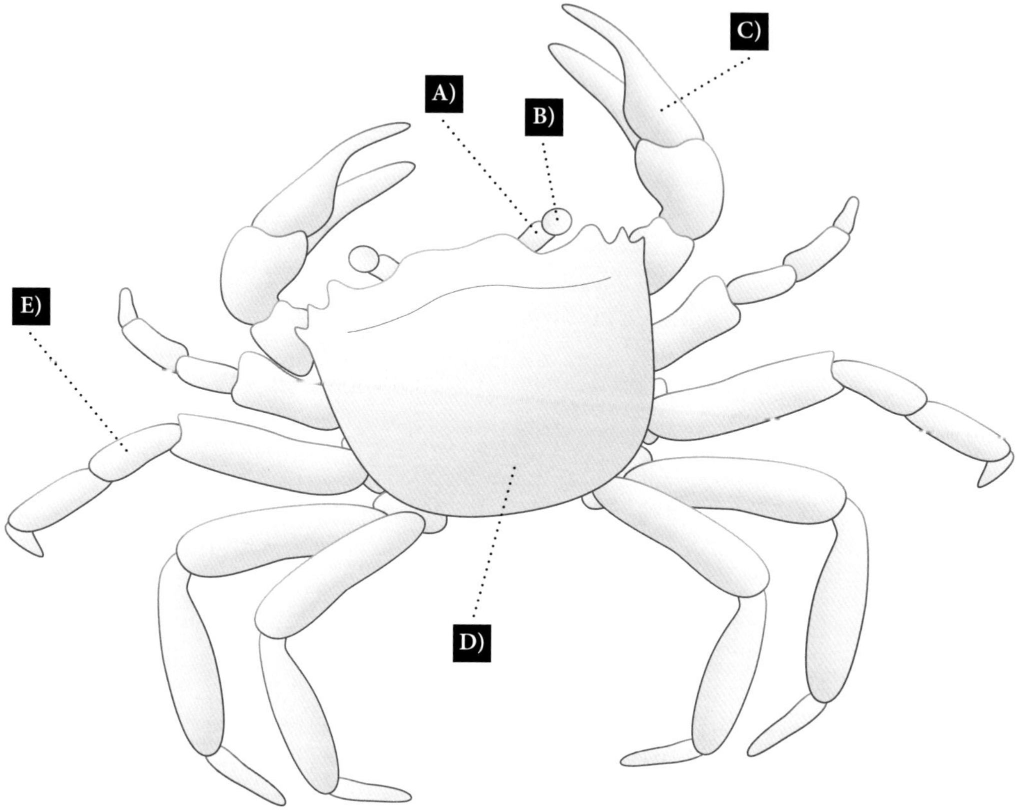


# ANATOMY

A) Eyestalk
B) Compound eye
C) Claw
D) Carapace
E) Walking leg

Female crab

Male crab

Keeping crabs is by no way easy. They are intelligent invertebrates with highly developed sensory organs and need special environments. As crabs have quite individual temperaments, their keeper needs to observe them closely to be able to meet their needs.

Crabs are highly interesting crustaceans to keep and can fill an aquaterrarium with life. However, they are quite susceptible to changing habitat conditions and to stress. They need lots of room and good hiding-places under and above the waterline. Many species require a land part as their life strategy is amphibian.

Differently from other crustaceans, the size of crabs is indicated by carapace diameter, not body length.

*Cardisoma guanhumi*, the Blue Land Crab, originates from subtropical coastal areas and is not easily available in trade.

# LAND CRABS

Right now, around 20 different land crab species (Gecarcinidae) are known. These crabs spend almost their entire life on firm ground. Only their larvae live in the sea for some time before they return to land after undergoing metamorphosis. After that point, they are only rarely found in water. They are distributed in the tropics, in coastal areas and on islands around most of the world. In their natural habitats, crabs play an important role in the ecosystem, as they burrow deep in the ground, helping to keep it aerated and preventing it from compacting, which greatly benefits local plant populations. Moreover, crabs eat carrion and dead leaves, which thus are returned to the nutrient cycle faster, and made available for plants again.

## *Cardisoma armatum*

Rainbow Land Crab, Tricolour Crab

*Cardisoma armatum* was probably one of the very first crabs available in the trade. This species inhabits mangrove woods, coastal alluvial land and river mouths near the sea. It is a semi-terrestrial species, which feels at home both on land and in water. The Tricolour Crab is considered very aggressive and does not tolerate the company of other crabs, even conspecifics, outside its mating season. An aquaterrarium needs to be sufficiently large, as this crab can live for 12 and more years provided that it is well cared for, and can thus grow to a considerable size. The land part of the aquaterrarium ought to take up 50 % of the total surface and should be covered in a loose substrate of sufficient depth so the crab can burrow. For reproduction, rainbow land crabs need brackish to marine conditions. The females release their larvae into the sea where they go through several stages and finally develop into young crabs, which leave the water after a few weeks to regain the coast.

In captivity, these crabs tend to eat anything they are offered. Their diet may contain fish and invert foods from the trade as well as unseasoned fish and raw unseasoned meat.

Body size: approx. 17 cm (6.7")
Temperature: 18 to 30 °C (64 to 86 °F)
Tank size: 100 cm (40") and more
Keeping conditions: single-housing
Area of distribution: West Africa
Reproduction: larvae need brackish to marine water

## *Gecarcinus lateralis*

Black-back Land Crab, Bermuda Land Crab, Moon Crab

*Gecarcinus lateralis* is a highly suitable crab for terraria. It tends to spend most of its time outside the water and only goes there to release its larvae, which makes these crabs totally unsuitable for aquaria. A shallow water bowl in their terrarium is sufficient, alongside high air humidity and frequent spraying. *G. lateralis* are not picky eaters and gladly accept foods of both vegetable as well as animal origin. A sufficiently large terrarium with a well-structured layout that provides many hideouts can house a small group (1 male, 3 females). These crabs have developed a sophisticated way of communicating with each other, passing on information through substrate-borne vibration signals, be it wooing a female or staking out one's territory against the claims of other males. The crabs produce these signals by rubbing movable body parts against each other. Another striking characteristic of *G. lateralis* is that freshly moulted crabs of this species do not inflate their new soft carapace with water, like most other crab species do, but, as they usually live in moist but not wet conditions, with the help of the gasses produced in their intestines.

Body size: approx. 10 cm (3.9")
Temperature: 18 to 28 °C (64 to 82 °F)
Tank size: 100 cm (40") and more
Keeping conditions: in pairs or small groups
Area of distribution: Florida, Central America
Reproduction: larvae need brackish to marine water

## *Gecarcinus quadratus*

Red Land Crab, Halloween Crab, Harlequin Land Crab

*Gecarcinus quadratus* is probably one of the most beautiful land crabs. The crabs of this species available in the trade usually originate from Costa Rica. Their importers give them various fanciful trivial names like "Halloween Crab" (note: this name is also used for *Gecarcinus ruricola*, amongst others) or "Purple Moon Crab", "Whitespot Land Crab", and many more. They need a large aquaterrarium with a big land part. As water part, a bowl of fresh water may suffice, given that the terrarium is kept moist.

In nature, these crabs are best observed by night, when they come out to eat. During the day, they tend to hide in deep burrows in order to avoid predators. Their staple diet mainly consists of dead leaves and tree seeds, which they carry into their burrows to eat in safety. *G. quadratus* is not a very aggressive crab provided that the aquaterrarium is sufficiently large and well-structured, and thus may be kept in small groups.

Body size: approx. 10 cm (3.9")
Temperature: 18 to 28 °C (64 to 82 °F)
Tank size: 100 cm (40") and more
Keeping conditions: in pairs or small groups
Area of distribution: Florida, Central America to Guyana
Reproduction: larvae need brackish to marine water

# MANGROVE CRABS

Mangrove forests are complex ecosystems, home to an incredibly large number of plants and animals perfectly adapted to the living conditions in these coastal areas and brackish river mouths. They are also called tidal forests, as they are periodically flooded following the rhythm of the tidal cycles. In the muddy areas of these forests, various mangrove crab species dig their burrows. These crabs play an important role in the ecosystem of these biotopes. They belong to various genera and often even to different taxonomical families. Some of them are known to climb trees, thus having found an efficient method of concealment from aquatic predators. Many mangrove forests have been obliterated by human activities, and for this reason as well as due to overfishing, mangrove crabs have become rare in some regions of the world.

Mangrove crab habitat on Cuba

CRABS

## *Pseudosesarma moeschi*

Red-Clawed Mangrove Crab

The red colour variety of *Pseudosesarma moeschi* is probably the most widely spread crab species in the aquarium hobby. However, there is also another colour variety with whitish-yellow claws, yellow and red claw fingers and brown-coloured body and legs. In general, the claws of adult males are larger and more colourful than those of the females.

These crabs need lots of room and love interaction with conspecifics (in smaller tanks, make sure there is only one male, though). The water part in the aquaterrarium may consist of brackish or fresh water. *P. moeschi* need lots of hideouts within and outside the water part. They require a cave to hide in especially after moulting. Crabs of this species can be co-housed very well with other crabs of comparable size. They are not picky eaters and accept most fish and invert foods. Make sure you feed a balanced diet, though.

After mating, the female retreats to a burrow, only to re-appear with a package of several hundred eggs under its apron after a few days. After three to four weeks she releases her larvae in water.

Breeding these crabs in captivity is possible yet quite complicated, as the larvae need a certain water salinity and the right food for growing well. *P. moeschi* can reach five to six years of age, given that they are well cared for. Like Fiddler Crabs and other crab species also inhabiting mangrove forests, they communicate by tapping the substrate and also by claw movements.

Body size: 3 cm (1.2")
Temperature: 20 to 30 °C (68 to 86 °F)
Tank size: 60 cm (24") and more
Keeping conditions: in small groups of three or more
Area of distribution: Thailand, Indonesia, Malaysia, Burma
Reproduction: the larvae need brackish to marine water

## *Pseudosesarma crassimanum*

Red Claw Crab, Spider Crab

Appearance and behaviour of *Pseudosesarma crassimanum* strongly resemble *P. moeschi*. However, there are some differences, mainly in the colour of the claw fingers, which are white and yellow in *P. crassimanum*. Whereas almost the entire claw with the exception of the fingers of *P. moeschi* is red, only the undersides of the claws of *P. crassimanum* are of a dark red. Males grow larger than the females and have larger claws, which they use for rivalry behaviour display and territorial disputes. The habitats are also comparable to those of *P. moeschi*, however, *P. crassimanum* can be found farther inland. Crabs of both species can be kept in one aquaterrarium without any problems.

In their native mangrove swamps, the crabs of this species dig short burrows in the loose ground. They play an important role in the ecosystem of tidelands and mangrove forests as their burrowing activity helps bring oxygen into the otherwise very cohesive soils. Their staple food consists mainly of dead leaves, which they carry into their burrows. Thus leaf decomposition is accelerated, and the ground is additionally enriched with nutrients.

Body size: 3 cm (1.2")
Temperature: 20 to 30 °C (68 to 86 °F)
Tank size: 60 cm (24") and more
Keeping conditions: in small groups of three or more
Area of distribution: Indonesia, Malaysia, Burma
Reproduction: larvae need brackish to marine water

## *Pseudosesarma bocourti*

Blue Thailand Crab

This crab species is rarely seen in the mangrove swamps along the coasts of South-East Asia as it is nocturnal and, in addition, rather shy. Even the smallest movement or the tiniest substrate-borne vibration sends it into its burrow.

A characteristic of the Blue Thailand Crab are its black claws with red tips. All in all, this species is quite peaceful in the aquaterrarium and can thus be kept in small groups – given that the tank provides sufficient room and places to hide. *P. bocourti* can be co-housed with other crab species of comparable size.

Since its main staple food in the wild consists of detritus and all kinds of plants and dead plant parts it is recommended to provide the Blue Thailand Crab mainly with food of vegetable origin. However, from time to time, black mosquito larvae, glassworms or bloodworms, earthworms or clam meat can be offered, too.

Body size: 4 cm (1.6")
Temperature: 20 to 30 °C (68 to 86 °F)
Tank size: 80 cm (32") and more
Keeping conditions: in groups of three
Area of distribution: Vietnam, Indonesia, Malaysia, Thailand
Reproduction: larvae need brackish to marine water

## *Sesarmops intermedium*

Flower Crab

In a well-structured aquaterrarium, *S. intermedium* may be kept in small groups. A good substrate for the water part is sand or fine-grained gravel, whereas the land part ought to consist of terrarium humus, which is less susceptible to mould. Both land and water parts ought to be structured by rocks or driftwood to offer the crabs a large number of hideouts. *S. intermedium* cannot do without them, so please make sure you provide these crabs with a sufficiently large number of places to hide.

These crabs are very peaceful with conspecifics and may even display social behaviour, especially when pairing off. When keeping *S. intermedium* in groups, make sure there are more females than males. After mating, the females may, depending on their size, carry well over one thousand eggs under their apron, and release the larvae after approximately four weeks.

Like most mangrove crabs, they are omnivores and, not being picky eaters, accept all kinds of food, as well as most vegetables and fruit like bananas, apples or pears. Rotting leaves, however, are among their favourites.

Body size: up to 4 cm (1.6")
Temperature: 20 to 30 °C (68 to 86 °F)
Tank size: 80 cm (32") and more
Keeping conditions: in pairs or groups of three
Area of distribution: Indonesia, China, Taiwan, Thailand, Japan and Korea
Reproduction: larvae need brackish to marine water

## *Neosarmatium africanum*

Spider Crab

For successfully keeping these small colourful crustaceans, a roomy aquaterrarium and several conspecifics are a must. Their social behaviour is highly sophisticated, and in their habitats, as many as 22 of these crabs have been found to live in one square metre, some of them adults, with the great majority being adolescents and juveniles. In addition to that, they are known to dig burrows that may be several metres in depth, which enlarges their habitat considerably. This large depth cannot be provided in an aquaterrarium, of course, this should however be made up for by giving them more room on the surface. These crabs should definitely be kept in groups.

*N. africanum* are less shy than most other crab species and often don't even hide if their keeper cleans and tidies up their aquaterrarium. They can even learn to take food from the hands of their keeper or a pair of tweezers. They will eat any and all plants in their aquaterrarium, and fish in the water part (brackish or fresh) are actively predated on.

Body size: 4 to 6 cm (1.6 to 2.3")
Temperature: 18 to 30 °C (64 to 86 °F)
Tank size: 80 cm (32") and more
Keeping conditions: in pairs or groups of three
Area of distribution: Eastern Africa from Somalia to South Africa, including the northwest of Madagascar
Reproduction: larvae need brackish to marine water

## *Metasesarma aubryi*

Red Apple Crab

This nocturnal crab inhabits coastal forests or mangrove areas. In the aquaterrarium it is not seen very often, especially not if the habitat is well-structured. Numerous rocks or driftwood, a sandy substrate and a small water part are sufficient. In nature, their staple food mainly consists of dead leaves and detritus, however, they also like to eat proteins of animal origin. In captivity, the entire range of fish and invert foods available in the trade can be fed.

*M. aubryi* only reproduce in brackish or marine waters, as the larvae have to undergo several stages and only return to land after their last metamorphosis. In the trade, berried females are sold sometimes, however, this crab species has not been raised successfully in captivity yet. Up to now it is not even clear which salinity the larvae need for successful development.

Body size: 3 cm (1.2")
Temperature: 20 to 30 °C (68 to 86 °F)
Tank size: 60 cm (24") and more
Keeping conditions: in trios or small groups of 5 to 6 crabs
Area of distribution: Indonesia, Thailand, Taiwan, Burma
Reproduction: larvae need brackish to marine water

## *Metasesarma obesum*

Batik Crab, Marbled Crab

The carapaces of Batik Crabs may vary greatly in colour, however, shades of black, cream and pink are present in almost any specimen. The aquaterrarium ought to have a mixed substrate consisting of sand and gravel, several rocks or driftwood under which the crabs can hide, and can be planted with various small grasses. Apparently, these crabs mainly feed on small beetles and other insects they find on the beach and close to rivers. Thus, they should also be provided with proteins of animal origin in captivity. In the terrarium trade, numerous small insects are available as food animals that may be offered to these crabs. In order to keep the aquaterrarium moist it should be sprayed with water at least once a day, and a small bowl of seawater should be provided. Depending on how well-structured the layout is, Marbled Crabs can be kept in small groups. Two males and four females seem to be the optimal stock for an aquaterrarium of a volume of 60 litres (15 US gal).

Body size: 2 to 3 cm (0.8 to 1.2")
Temperature: 20 to 30 °C (68 to 86 °F)
Tank size: 60 cm (24") and more
Keeping conditions: in small groups of 5 to 6 crabs
Area of distribution: Pacific part of Madagascar, East Africa, Indonesia, Australia to Taiwan
Reproduction: larvae need brackish to marine water

## *Perisesarma huzardi*
African Mangrove Crab

When observing these crabs in their natural habitat during low tide you will soon see that they are omnivorous and try to eat just about anything that happens to end up in their claws. In the aquaterrarium they can be fed vegetables and fruit and all sorts of fish and invert food available in the trade.

These crabs need an aquaterrarium with a length of at least 80 to 90 centimetres (32 to 35"), with a large land part and several pieces of driftwood. Larger females may carry up to 3.5 million eggs. When they hatch, the larvae are released into the sea, where they go through several larval stages and return to the tidal zone when having metamorphosised into crabs. The water part in the aquaterrarium may contain fresh or marine water. Co-housing with crabs of a comparable size of the genera *Chiromantes*, *Neosarmatium* or *Sesarmops* is usually possible without any problems, provided that there is sufficient room, of course.

Mangrove habitat

Body size: 4 to 6 cm (1.6 to 2.3")
Temperature: 20 to 30 °C (68 to 86 °F)
Tank size: 80 cm (32") and more
Keeping conditions: in pairs or groups of three
Area of distribution: West Africa
Reproduction: larvae need brackish to marine water

# DWARF LAND CRABS

The crabs of the genus *Geosesarm*a belong to the sesarmid family. They were first imported into Europe in 2006, and right now they are the most popular crabs in the trade. The genus *Geosesarm*a is widely distributed in South-East Asia. To date, 49 species belonging to that genus have been scientifically described. Most of these small crabs have striking colours, and they are fully grown with a carapace width of only one to two centimetres (0.4 to 0.8"). These terrestrial crabs are only rarely found in water. This is the reason why they are absolutely not suitable for an aquarium without a land part. The natural habitats of *Geosesarma* are located near waterbodies, mainly small rivulets or mere trickles of water, along whose edges they dig their burrows.

## The aquaterrarium

A dedicated aquaterrarium for dwarf land crabs can be decorated with mosses, rocks and driftwood. The air humidity should be as high as possible. Some *Geosesarma* species are highly endangered in their countries of origin as their areas of distribution are limited to small forest zones that are in danger of being lumbered or drained. The population density of these crabs in their natural habitats is low, and the males stake out territories, which they defend against intruders. Please keep this in mind when stocking your aquaterrarium: Either put in only one male and two females, or at most a small group of five to six crabs – two males and three to four females, given that the aquaterrarium is sufficiently large.

Different species should not be co-housed, as the more dominant species will attack and even kill the weaker crabs. In general, female dwarf land crabs carry relatively few but rather large eggs, from which fully developed young crabs will hatch after up to four months. However, *Geosesarma maculatum* and *Geosesarma perracae* are known to release larvae into fresh water, which develop into young crabs after a short time. Separating the offspring from their parents is highly recommended when raising them in captivity, as the adults will predate on them, and in the relative confinement of the aquaterrarium the young crabs will have difficulty in avoiding them. Most species are carnivorous, hunting above as well as below the waterline. They can be offered live house crickets or other insects on the land part, in fact, any live food from the terrarium trade will do nicely. Under water they have been observed to predate on small fish or shrimp and other small inverts.

Even though some of these species look very much alike, there are huge differences regarding behaviour, especially in reproduction and brood care. Some of them reproduce exclusively on firm ground, others need fresh water, some undergo direct development, i. e. fully developed young crabs hatch from the eggs, others belong to the unspecialised type whose larvae require water and most frequently even marine conditions to develop.

## *Geosesarma dennerle*
Vampire Crab

The exact origins of Vampire Crabs were in the dark for a very long time, however, in mid-2011 we travelled to Indonesia where we were able to find their habitat. As it turned out there are at least five different species from various regions that are currently imported under the common name "Vampire Crab". The species shown has recently been described as *Geosesarma dennerle*. It was found near Cilacap on Java/Indonesia. The rivulet where these crabs live is approximately two to three metres wide. The adult crabs apparently don't burrow very much, as we found all the crabs under rocks or in dense vegetation save a single one that was sitting in the water beside a stone. A striking characteristic of this biotope is the great number of insects sharing their habitat with the crabs, probably serving them as food. Juvenile crabs are mainly found on the edges of the small trickles of water that feed the rivulet, where they dig short burrows into the soft ground. No adult crabs can be seen in the open during the day, however, we observed several young crabs near the creek. In the aquaterrarium these crabs should be kept in a small group in a dedicated tank. They should not be co-housed with other species of the genus *Geosesarma*.

Body size: 2 to 2.2 cm (0.8 to 0.9")
Temperature: 20 to 27 °C (68 to 80 °F)
Tank size: 60 cm (24") and more
Keeping conditions: in small groups of 5 to 6 crabs
Area of distribution: Indonesia
Reproduction: in freshwater

Vampire Crab habitat on the island of Java

## *Geosesarma hagen*

Red Devil Crab, Red Vampire Crab

We found the Red Vampire Crab in a small moist forested area near Cilacap in the south of Java. They can be seen out in the open by day and night. These crabs mainly live on and in the small trickles of water that run down the mountain sides, where they burrow in the soft loamy ground. They eat small insects and other invertebrates, and sometimes even their own offspring.

The Red Vampire Crab ought to be provided with a well-structured aquaterrarium containing a small water part, and it needs a high air humidity. As these crabs are basically carnivores they should be offered live food from terraristics, any of which they gladly accept. When well-cared for, these crabs can grow two to three years old at least, and they seem to get used to their keeper after a rather short time, which allows feeding them by hand.

Body size: 2 cm (0.8") (0.8")
Temperature: 20 to 26 °C (68 to 70 °F)
Tank size: 60 cm (24") and more
Keeping conditions: in small groups of 5 to 6 crabs
Area of distribution: Java, Indonesia
Reproduction: in freshwater

## *Geosesarma* sp. "Blue"

Blue Vampire Crab

Another nocturnal crab, *Geosesarma* sp. "Blue", is also sold as Blue Vampire Crab in the trade. Their place of origin is not yet exactly known, however, we assume they might come from Indonesia, as they are shipped to Europe and Japan from Jakarta, like most crabs of the genus *Geosesarma*. Apparently, the Blue Vampire Crab prefers to dig short burrows in soft ground, and thus the substrate in their aquaterrarium ought to consist of a loose mixture of sand and terrarium humus.

Body size: 2 cm (0.8")
Temperature: 20 to 27 °C (68 to 80 °F)
Tank size: 60 cm (24") and more
Keeping conditions: in small groups of 5 to 6 crabs
Area of distribution: Indonesia
Reproduction: in freshwater

## *Geosesarma* cf. *tiomanicum*

Orange Vampire Crab, Disco Vampire Crab

Not fully identified yet, the Orange Vampire Crab still remains a bit of a mystery. Its exact origins continue to be in the dark, as most Indonesian collectors tend to protect their territory very closely. Thus we do not know much about their habitat and their life strategies, however, these crabs presumably live near flowing waters, and their behaviour and eating habits should not differ too much from those of other *Geosesarma* species sold in the pet trade.

A special characteristic of the Orange Vampire Crab is that it can change its body colour, depending on light intensity. Under strong light these crabs assume an intensive orange colour, their claws turn purple or pink. However, if the aquaterrarium is rather dark their basic colour is a blueish grey. The females are berried for an exceptionally long period of time. Mengedoht & Rademacher observed in one of their aquaterraria that the young crabs only left their mother for good after as long as 214 days.

Body size: 2 cm (0.8")
Temperature: 20 to 26 °C (68 to 79 °F)
Tank size: 60 cm (24") and more
Keeping conditions: in small groups of 5 to 6 crabs
Area of distribution: Java, Indonesia
Reproduction: in freshwater

## *Geosesarma notophorum*
### Mandarin Crab

The Mandarin Crab is a rather peaceful species that is best kept in a dedicated terrarium. It is a nocturnal species and hides during the day, mostly in short burrows it digs in the soft ground, or under dead leaves, rocks or driftwood. Mandarin Crabs are known to be excellent climbers, and thus their aquaterrarium ought to be well-structured not only on the ground but also vertically. It may be somewhat higher, too. If their aquaterrarium is constantly kept moist, these crabs do not necessarily need a water part. If provided with one, they have been observed to use it sparingly and only for a very short time.

The females apparently do not eat at all when berried, and when the juveniles have hatched they carry them on their backs for several days in a row before releasing them for good. This special brood care behaviour has only been observed in this species so far. Mengedoht & Rademacher observed that the females carry their eggs for approximately 116 days. As soon as the young crabs have left their mother it is highly recommended to raise them in an aquaterrarium or container of their own in order to prevent them from being eaten by their own parents. The young crabs have been observed to eat each other, too, and thus not all of them reach adulthood. *Geosesarma* cf. *notophorum* "White Hand" is also sold in the pet trade. Currently it is unknown if these are only a colour variant of the species, or belong to a new species in their own right.

Body size: 1 to 1.5 cm (0.4 to 0.6")
Temperature: 20 to 26 °C (68 to 79 °F)
Tank size: 30 cm (12") and more
Keeping conditions: in trios.
Terrestrial crabs, high air humidity is sufficient.
Area of distribution: Pulau Lingga, Indonesia
Reproduction: in freshwater

## *Geosesarma krathing*
### Tangerine-Head Crab

The crabs of the species *Geosesarma krathing* are mainly nocturnal and, in addition, very furtive. In their natural habitat they are usually found very close to water, where they tend to sit in the shallow zones near the edges, but they have also been observed to climb smaller bushes. According to scientific literature, up to seven crabs have been found per square metre. They mainly feed on detritus, but also on insects. The females carry an average of 50 eggs, from which fully developed young crabs hatch. During their brooding phase, the females are hardly ever seen out in the open as they retreat into the burrows they dig. The young crabs need a water part for successful development. Adult crabs use the water as well as the land part for their activities. This species is considered to be relatively peaceful, however, a dedicated aquaterrarium is highly recommendable.

Body size: 1 to 1.5 cm (0.4 to 0.6")
Temperature: 20 to 26 °C (68 to 79 °F)
Tank size: 30 cm (12") and more
Keeping conditions: in trios
Area of distribution: Thailand
Reproduction: in freshwater

# SULAWESI CRABS

Together with the colourful shrimp from Sulawesi that were brought onto the market several years ago, some crab species have also been imported. Those species originating from the Ancient Lakes are most frequently found in the trade as the collectors catch them together with the shrimp. There are three different types of crabs in the lakes themselves. The unspecialised species of the genus *Parathelphusa* are found in the feeder rivers of the lakes as well as in the lakes. The detritus-eating crabs of the genera *Migmathelphusa* and *Nautilothelphusa* are characterised by smaller claws and their peculiar hind legs that have evolved in a way that enables them to dig into loose sediment very quickly. The snail-eating crabs of the genus *Syntripsa* are the largest species in the Ancient Lakes of Sulawesi, where they are endemic, like the others.

## *Parathelphusa ferruginea*
Towuti Crab

Towuti Crabs are mainly found under water; they only leave it occasionally and are rarely seen on land. However, they still need an opportunity to leave the water when kept in captivity. The first descriptive authors of this species assumed there are three colour variants with differently coloured

legs. However, aquarium-kept *Parathelphusa ferruginea* have been observed to be able to change their colour very quickly, thus it is possible that the different colours described might not be variants but mood-related.

In the Ancient Lakes, these crabs mainly eat snails and other small inverts they crack open with their strong claws. When kept in an aquarium, they ought to be offered clam meat or other food rich in proteins. Young crabs are relatively peaceful towards their conspecifics, however, the adults are quite incompatible with each other. A pair can only be kept in sufficiently large tanks with a very large number of hideouts.

Body size: 5 to 6 cm (2 to 2.3")
Temperature: 26 to 30 °C (79 to 86 °F)
pH value: 7.5 to 8.5
Tank size: 100 cm (40") and more
Keeping conditions: in pairs
Area of distribution: Lake Towuti and Lake Mahalona, Sulawesi
Reproduction: in freshwater

## *Parathelphusa pantherina*

### Panther Crab

In its natural habitat, the Panther Crab is almost always found in water, however, from time to time it can also be observed to be sitting under roots, driftwood or rocks on the lakeshore. In nature, these crabs eat detritus, snails and fish, they usually do not munch on green plants. As they grow to a considerable size they need a correspondingly large tank. Even though they lead an aquatic life for most of the time we recommend providing them with a seat above the waterline nonetheless. A rock or a piece of driftwood ought to suffice. Please make sure the aquarium is sealed so they cannot escape, as they do not survive on dry land for a long period of time. If you plan on keeping a pair or group of three in your aquarium, please make absolutely sure that there is at least one hideout per crab. Adult crabs display aggressive behaviour towards conspecifics, and they predate heavily on fish. *Parathelphusa pantherina* should definitely not be co-housed with other crab species. They do reproduce in freshwater, however, they have not been reported to breed successfully in captivity. These crabs can reach an age of up to ten years, and those available in the trade are mostly juveniles.

Body size: 8 to 9 cm (3.1 to 3.5")
Temperature: 26 to 30 °C (79 to 86 °F)
pH value: 7.5 to 8.5
Tank size: 120 cm (48") and more
Keeping conditions: single
Area of distribution: Lake Matano, Sulawesi
Reproduction: in freshwater

*Syntripsa matanensis* is a specialised snail-eating crab that sometimes is imported under the common name "Purple Panther Crab".

In Lake Towuti and Lake Mahalona, *Syntripsa flavichela* is found, also called "White Claw Crab".

# MINI CRAB

***Limnopilos naiyanetri***
Micro Crab

The smallest crab available in the trade lives exclusively in fresh water. Up until now they have never been observed to leave the water at all. In their natural habitat, they are found in slow-flowing zones of small to large rivers, where they find abundant food and hideouts in the tangled roots of the trees on the riverbanks that reach into the water, and they have also been observed to be sitting in the roots of diverse floating plants. In the aquarium, they ought to be provided with finely ramified driftwood and/or floating plants. They reproduce in freshwater. The females carry eggs frequently and at regular intervals, however, raising their offspring is very difficult and has not successfully been done in captivity yet.

Body size: 1 cm (0.4")
Temperature: 24 to 28 °C (75 to 82 °F)
pH value: 7.5 to 8.5
Tank size: 10 litres (2.6 US gallons) and more
Keeping conditions: in groups of 5 to 6 crabs
Area of distribution: Thailand
Reproduction: in freshwater

Dense Fiddler Crab population in Argentine.

**FIDDLER CRABS** play a crucial role in the ecological balance of tidelands and mangrove forests as they have the important task of aerating the ground in coastal swamps and marshland. Their burrowing activities are very important for the well-being and healthy growth of grasses and other coastal or marsh plants. Moreover, Fiddler Crabs are a good indicator animal for environmental pollution, as they are highly susceptible to polluted water and insecticides. Crabs of the genus *Uca*, which comprises around 100 scientifically described species, lead a semi-terrestrial life and inhabit burrows they dig in the sandy or muddy ground at the waterside of tropical or subtropical coasts. Some of these species are really beautiful in colour, and especially children in the tropics love catching them for fun. If you are lucky and manage to catch one of these crabs you might find that some of these species can change their colour really fast and fade from spectacular to very inconspicuous. The colours of some species will also fade during nighttime, when temperatures are cooler and it is dark.

The most striking characteristic of male Fiddler Crabs are their extremely conspicuous claws, one of which tends to be considerably larger and of a strikingly intense colour. It may amount to as much as 75 % of the total weight of the crab. It is a warning signal to male competitors that cannot be overlooked, and at the same time it serves to woo the females, whose claws are smaller and do not differ in size. The large claw of the male is not very strong and thus cannot be used in a fight or as a tool to crack open clams or snail shells, its only purpose is display. Male Fiddler Crabs may be left- or right-handed, and they can even change their dexterity if the large claw has been lost in a fight, for example.

*Uca stenodactylus*

Fiddler Crabs always live in a tunnel or even a tunnel system they dig themselves. They retreat to this burrow during high tide. You will never find a Fiddler Crab that has ventured into free water voluntarily. The burrows are closed off against the water with sand or mud. Some species seal them off so tightly that no water can enter at all and that some air always remains in the tunnel, which they desperately need for breathing. Many Fiddler Crab species would drown under water as their gills have evolved in a way that allows them to breathe atmospheric air only.

During breeding time in spring to summer, many species synchronise their lives exactly with the tidal activity. The females are ready for mating approximately every four weeks, and they always lay their eggs during the spring tides, which occur when the moon is full or new. Then the differences between high and low tide are greatest. During these times, the males dig larger tunnel systems than usual, where copulation takes place. These burrows can be 30 to 40 cm (12 to 16") deep, and the males defend them fiercely against any competitors.

After high tide, the females leave their burrows and go searching for a mate, which they do meticulously, strictly assessing the males. Female Fiddler Crabs are very picky when it comes to choosing a mate, and thus the males have to go to great lengths in courtship. They wave their large claws without pausing, and whenever a female shows interest and stops in front of a male, he starts waving his large claw with even more intent. If the female does not approach further the male starts running to and fro between her and his burrow to show the lady of his heart the way and to lead her into his tunnel sys-

Territorial struggle

tem. If the female likes what she sees she approaches his burrow but stops again at the entrance. The male crawls a short way into the burrow and starts drumming the tunnel walls with his large claw. The female perceives the vibrations the male thus creates and – if she likes what she feels – follows him into the burrow. The male leads the female into a cave at the end of the tunnel, turns around immediately afterwards and closes off the entrance with sand or mud.

After mating, the male leaves the burrow and recloses it, only to dig a new one to get a chance to mate with another female. He repeats this activity until his reserves are exhausted. The female stays protected in the closed-off brooding burrow, where the eggs under her apron can develop in safety.

When the larvae inside the eggs are fully developed, the female leaves the burrow and releases them into the sea. She chooses the exact point in time when the tide is highest with astonishing precision, and thus the tiny larvae are carried out into the open sea with the ebbing water of the spring tide.

Female next to a burrow

## Keeping Fiddler Crabs in an aquaterrarium

Fiddler Crabs may vary a lot in form and coloration even within a species, and thus photos often may not suffice for exact identification. However, the large claws of the males have clearly different forms in different species (to avoid misunderstandings and interspecific mating), thus the claw form is a good characteristic for species identification.

*Uca annulipes*

Those crabs are interesting and colourful crustaceans that are sometimes sold as aquarium pets in the trade. However, they should be kept in an aquaterrarium rather than an aquarium as they need a large land part consisting of sand, and a water part that may hold fresh, brackish or marine water.

It is most recommendable to keep only a few males and a larger number of females. Otherwise, if the aquaterrarium is too small the males might start fighting over their territories, even though in nature, Fiddler Crabs live in colonies of many individuals and are thus used to high population densities. Not much interspecific aggression can be observed either. If a male happens to lose a claw in a fight it is replaced during the next moult. It is a little smaller at first but grows every time the male sheds its skin. Grown Fiddler Crabs moult once or twice a year.

Fiddler Crabs can in no way be kept in a marine coral tank. If they are kept submersed all the time they will constantly try to flee the tank as they are in acute danger of drowning. In many cases, they succeed but will die somewhere in the room that houses the aquarium if their keeper doesn't find them soon. Fiddler Crabs may be able to survive enormous losses of fluid, however, indoor air is simply too dry. If a crab is missing you could try putting a shallow bowl of water on the floor, or even a moist sponge or piece of cloth to lure it, as Fiddler Crabs like their surroundings moist.

The ground in the aquaterrarium ought to consist of at least a 10 cm (4") thick layer of sand or mud (which can be mixed with gravel or granulated aquarium substrate) so the crabs can burrow.

The crabs need to be provided with a water part, too, with fresh, brackish or marine water, depending on the species you are planning to keep, or all three of them so the crabs can choose the water they prefer. Fiddler Crabs often dig their tunnels under rocks and driftwood for additional protection. The burrow is an important factor when it comes to reproduction as the female picks the male by its quality. Thus the aquaterrarium ground is crucial, as many species are specialised in digging their tunnels in certain substrates (sand, mud, or a mixture of both). If these prerequisites are met, the crabs may even mate in captivity, and the females may carry eggs.

In most cases, however, it may be too difficult to raise the offspring. The first larval stage, the so-called zoea, need open-sea conditions for survival. If you give the berried female a marine water tank to release her larvae in her eggs may even hatch, however, the larvae need to go through different stages until they develop into adult Fiddler Crabs, and these larval stages are difficult to care for in an aquaterrarium.

Depending on the species, these crabs can reach an age of two to three years, however, in captivity they often die after one or two years. When kept in near-perfect conditions they may reach an age of over six years, though. Fiddler Crabs and fish should not be socialised, even though these crabs do not predate on fish or other larger animals but mostly eat detritus and organic material they find in the mud on the coast. However, if fish try to attack the crabs or are just curious the crabs will try to defend themselves, and males might even use their large claw for doing so.

*Uca vocans*

*Uca uruguayensis*

*Uca arcuata*

*Uca tangeri*

# FIDDLER CRABS

*Uca annulipes*

*Uca oerstedi*

*Uca vomeris*

*Uca terpsichores*

*Uca crassipes*

*Uca stylifera*

# FIDDLER CRABS

*Uca lactea perplexa*

*Uca pugilator*

*Uca stenodactylus*

# SNAILS

# SNAILS

## 27 SPECIES

Snails belong to the molluscs and have been assigned to the class of Gastropoda (gastropods) – second only to arthropods when it comes to species diversity in the animal kingdom. In fresh water, you will exclusively find snails with shells. Aquatic slugs, or nudibranchs, live in saltwater only.

A freshwater snail moves on its muscular foot, which it can fully withdraw into its shell. Some species have a trapdoor-like lid they use to close off their aperture in order to protect themselves from predators or adverse environmental conditions, like drought, for example. The shell is built by secretions of the visceral sac epithelium and the mantle rim on the snail's back and mainly consists of calcium carbonate.

Some snail species are pulmonate and thus have to go to the water's surface to breathe air, others have gills and can therefore live under water permanently. The vast majority of freshwater snails breed in freshwater, however, there are several species living in fresh water habitats that need brackish or even marine conditions for reproduction, as they go through several pelagic (i. e. free-floating) larval stages during their development from egg to adult snail. True freshwater snails, on the contrary, have fully developed offspring. In the hobby, representatives of both reproductive strategies are popular. Almost 10 000 different freshwater snail species have been described. However, this group surely will have to undergo a thorough revision – the genera and species are in dire need of re-classification.

## APPLE SNAILS

Among the most decorative snails suitable for aquaria, and distributed in many tropical regions in the world: Apple Snails inhabit very different ecosystems. For breathing they have developed an interesting combination of lungs and gills perfectly adapted to their requirements. Many Apple Snail species, especially those of the South-American genus *Pomacea*, have an anatomical characteristic: the siphon (a kind of tube) they use for breathing. This snorkel-like appendage allows them to remain below the water's surface while breathing air.

Another Apple Snail characteristic is the way they reproduce: Most species lay their eggs in hard clutches on plant parts or other objects above the waterline. In contrast to many other snails, Apple Snails are not hermaphroditic but dioecious, i.e. they have separated sexes. Thus, to have offspring, you need a male and a female snail. Unfortunately, sexing them is difficult. The sex of a live snail can only safely be determined when you see it lay eggs or during mating. Many Apple Snail species are voracious plant eaters and need a sufficient amount of vegetable food or else they will turn to your aquarium plants. The only exception here are *Pomacea diffusa*. This species can literally starve to death amid green vegetation if not fed otherwise.

It is necessary to cover your tank tightly if you want to keep Apple Snails in order to prevent them from going on nightly tours. They do, however, require some inches of space between the water surface and the lid to make sure they can breathe air easily. Apple Snails are not finicky regarding the water quality, however, like all snails they require calcium to build up their shell. If the water is too soft (and thus contains little calcium) you might want to increase their calcium intake by adding limestone, seashells or special preparations from the pet trade. Apple Snails are not picky eaters. They will eat just about anything as long as they can rip small chunks off with their radula (rasping tongue) and bring them into their digestive tract. Vegetables like cucumbers, spinach, carrots and lettuce, plus fish food, dead fish, other snails as well as their egg clutches, algae, sea shrimp – all these are part of the diet of these extremely voracious snails.

As the bowels of Apple Snails contain large numbers of microorganisms they need for digesting their food, the water of a tank containing these snails might become slightly turbid in time, as some of these microorganisms are excreted with the snail's faeces. However, this effect is not detrimental, at the most it is a little annoying for the onlooker. In some cases, these microorganisms may even serve as food for very small fish larvae. However, if you would rather have clear tank water, a good filter might help.

As Applesnails originate from the tropics, temperatures in their tank ought to be between 18 and 30 °C (64 to 86 °F). With increasing temperatures, the activity of these snails increases. At 18 °C (64 °F) they are almost inactive, whereas they are quite agile and move around a lot at 24 °C (75 °F) and above. However, temperatures do not only influence their activity but also how long they live and how fast they reproduce. Higher temperatures make for shorter life cycles, and the snails will die earlier. At lower temperatures they might live for around four years.

Clutch of *Pomacea* sp. on a large *Echinodorus*, and on the trunk of a tropical tree in Rio Sucuri, Brazil

## *Asolene spixi*

Zebra Apple Snail, Spixi Snail

Easy to discern from other snail species thanks to its striking colours and pattern, unfortunately not to be found in the pet trade very often, the shell of *Asolene spixi* is yellow or cream-coloured with clear black stripes, its body is of a yellowish brown with dark spots or dots. The Zebra Apple Snail is dioecious. Fertilised females attach their egg clutches to water plants or other objects under water. There are up to 50 eggs per clutch. Depending on the water temperature the young snails need three to four weeks to hatch. It is impossible to sex these snails from the outside, so keeping a small group is recommended to make sure you have both sexes. Zebra Apple Snails are known to eat aquatic plants even though they have feeding alternatives. Experience shows that plants such as Waterweed (*Egeria densa*) are not eaten at all. *Egeria densa* is illegal in some countries, though.

Size: 2 to 4 cm (0.8 to 1.6")
Temperature: 15 to 26 °C (59 to 79 °F)
pH: 7.0 to 8.5
Tank size: 20 litres (5 US gal) and more
Distribution: South America, from southeastern Brazil to northern Argentina

## *Marisa cornuarietis*

Giant Ramshorn Snail

At first glance, *Marisa cornuarietis* does not look like an Apple Snail at all with its ramshorn-shaped shell. Up until now, only two species of the genus *Marisa* have been described. They are hard to discern as they look quite similar. However, in the trade we only have *M. cornuarietis*. It is known as a plant eater voracious enough to rid a planted tank of all greenery in no time. Only waterweed and a few other plant species are not a part of its diet. Giant Ramshorns also eat brush algae and fish eggs and rasp off small chips of wood and brown tree leaves, which are a good dietary supplement.

*Marisa cornuarietis* is dioecious. Its egg clutches may contain as many as 200 (!) eggs and are laid below, but close to the waterline on plants or driftwood. Depending on the water temperature, the young snails hatch after two to three weeks. Characteristically, the shell of *M. cornuarietis* is basically flat and on the one side it is yellow with dark stripes, on the other side it is simply brown. The body is yellow to whitish and has a dark spotted pattern.

Size: 4 to 6 cm (1.6 to 2.4")
Temperature: 18 to 30 °C (64 to 86 °F)
pH: 7.0 to 8.5
Tank size: 50 litres (13 US gal) and more
Distribution: South America (e.g. Brazil), Central America, USA (Florida)

## *Pomacea diffusa*

### Spike-Topped Apple Snail

Until recently this snail was known as *Pomacea bridgesii* in the hobby, however, examinations have actually identified it – and all its colour variants, too – as *Pomacea diffusa*. Its wild form, not often found in the hobby, is of a rather inconspicuous brown with a dark striped pattern and a light-coloured body. Its white-footed yellow variant is far more widely spread, as are specimens with blue and purple shells with light or dark-coloured bodies. All these colour forms interbreed. The females lay clutches containing 200 to 400 eggs above the waterline. Most of the offspring survive given that the conditions in the breeding tank are favourable. In an aquarium, this snail will eat soft or dead plants and is best fed with pelleted fish food, boiled vegetables or deep-frozen spinach. They also like blanched vegetables. Please make sure to feed amounts the snails can eat within a few hours so the water quality is not negatively affected.

*Pomacea diffusa* do not eat healthy aquarium plants given that the snails find enough food, which makes it a suitable snail for planted tanks, too. The Spike-Topped Apple Snail is dioecious and reaches adulthood after four to six months. The females lay their egg clutches above the waterline on plant stems, wood or other firm objects like, for example, a tank cover. These clutches are light in colour, cream-coloured or pinkish and can thus be discerned from those of other Apple Snail species rather well.

Size: 4 to 5 cm (1.6 to 1.9")
Temperature: 18 to 30 °C (64 to 86 °F)
pH: 6.5 to 8.5
Tank size: 60 litres (15 US gal) and more
Distribution: Brazil, Peru, Bolivia

Unfortunately, the EU has prohibited trade with all snails of the genus *Pomacea*. The reason for this law lies in an increased appearance of *Pomacea canaliculata* and *Pomacea insularum* in the Ebro Delta in Spain, where these snails are said to damage plants. This prohibition applies to all snails of the genus *Pomacea*, however, amongst which you can find the Spike-Topped Apple Snail (*Pomacea diffusa*), the Golden Apple Snail (*Pomacea canaliculata*) and the Island Apple Snail (*Pomacea insularum*). The ban on distributing Apple Snails also applies to hobbyist snail keepers. Although the possession and the breeding of such snails is not prohibited, they must not be sold or given to other hobbyists. The Apple Snails *Asolene spixi* and *Marisa cornuarietis* remain unaffected by this rather questionable decision, though.

# NERITES

Most representatives of this family live in the sea, however, there are some species that left their original marine habitats independently from each other and are now found in estuaries of rivers and creeks. They have attractive patterns and unique shell forms, they are relatively robust and undemanding, and moreover, they are good algae eaters. All of these factors make them highly coveted in the hobby.

*Clithon diadema*

It is quite difficult to identify the exact species of a Nerite as their outward characteristics, like pattern and shell colour, may vary greatly. Moreover, pattern and colour tend to change when the snail comes to live in different conditions. The intraspecific variability is thus enormous. We have been able to photograph at least thirty different patterns and shell colours of *Clithon diadema*. Even renowned experts often disagree when it comes to determining characteristics, which has frequently led to different conclusions regarding species affiliation in scientific and aquaristic literature. The systematics of this entire family have not yet been fully examined. Currently, the number of different Nerite species is estimated to lie between 50 and 175, as many species are rather difficult to distinguish. Practically all Nerites belong to the primitive reproduction type and only reproduce in brackish or marine waters. Their egg cocoons can be found on hard substrates like rocks, wood or even the aquarium glass, depending on the species. After a few weeks or months, depending on water temperatures, water depth or other parameters, the larvae hatch. In nature, they drift to the sea with the current. There they feed on microscopic unicellular organisms. However, in a freshwater tank, this is impossible, and the larvae are condemned to die off after a short time.

There are only a few species of the genus *Theodoxus* (River Nerites) that belong to the specialised type. However, these snails are not often found in the hobby. The Nerites available in the trade can live as long as several years. In freshwater tanks, *Neritina turrita* has been known to live for six to eight years provided they get good care. Most of these snails have already reached adulthood when they are introduced to our tanks, which makes it difficult to say anything definite about their true life expectancy, though. They do not demand special water parameters or temperatures, however, they need a sufficient amount of calcium. When living in soft water their already slow growth is stunted even further. Substrates containing calcium (e.g. empty snail shells or cuttlebone, limestone rocks, sea shells or special preparations from the pet trade) may help provide them with the sufficient amounts.

## *Clithon subgranosus*
### Horned Nerite Snail, Sun Snail

This eye-catching snail species is another one to originate from the tidal zones of rivers and creeks. Here you can find it on rocks and driftwood. During low tide, they live in pure fresh water, during high tide their habitat is immersed in brackish water, so they undergo constant changes in salinity.

In the aquarium, they are relatively undemanding, however, they dislike temperatures of under 20 °C (68 °F). As with other species of the genus *Clithon*, the Sun Snail is very variable in form and colour. There are specimens with large horns, and others completely without them. In the aquarium, these horns do not form at all, as it seems. The exact reason why they grow is still very much in the dark, it could have something to do with the tides.

The Sun Snail is dioecious, however, the larvae need marine conditions to develop.

Size: 1 to 1.5 cm (0.4 to 0.6")
Temperature: 20 to 30 °C (68 to 86 °F)
pH: 7.0 to 8.5
Tank size: 10 litres (2.6 US gal) and more
Distribution: Solomon Islands and the Philippines

## *Neritina pulligera*
### Dusky Nerite, Baseball Hat Nerite

Also called Diamond Snail or Red Lips Nerite. Almost every aquaristic publication seems to use another trivial name for this species. *Neritina pulligera* has proven to be quite robust in the aquarium, and

it is a very good algae eater that even goes after brush algae. These snails are dioecious, but do not reproduce in a freshwater tank, even though they lay egg cocoons, mostly on hard substrates and on the shells of other snails. In their wild habitat, their larvae drift into the sea with the current of the rivers or brooks they hatched in. There they develop into snails, which may take weeks. The young snails first live in estuaries and then start wandering upstream, back into freshwater habitats. This is almost impossible to mimic in the hobby. *Neritina pulligera* is very widely distributed in nature, the reason for which is the long time the larvae spend in marine conditions while developing where they may drift long distances with the current. During the time we had them in our aquarium they did not leave the tank, however, a tank cover is recommended for keeping this species, too, in order to avoid any unpleasant surprises.

Size: 2.5 to 3 cm (1 to 1.2")
Temperature: 18 to 30 °C (64 to 86 °F)
pH: 7.0 to 8.5
Tank size: 80 litres (21 US gal) and more
Distribution: Western Indo-Pacific region to Taiwan, South Africa to Madagascar

## *Clithon diadema*

### Spiky Nerite, Bumblebee Nerite

The Bumblebee or Spiky Nerite from the Indo-Pacific region is quite variable. These snails are most frequently found in flowing and only rarely in stagnant brackish waters of river estuaries on rocks or driftwood, where they are subject to rapidly changing water conditions. At high tide, brackish water streams into the estuary, and when the tide is low, fresh water prevails. However, these Nerites can permanently live in fresh water, which makes them highly suitable for nano tanks and small aquaria. Patterns and colours may vary strongly in this species, however, in the trade you typically find the yellow and black striped variant with very attractive large spikes on the upper side of the shell. In the aquarium, these snails do not grow any more spikes, thus it may be safely assumed that this special growth pattern has something to do with the tides. The colour variant described above is probably collected in one region only, which explains why we do not have more variants entering the trade. The Spiky Nerite does not reproduce in freshwater and does not tolerate low temperatures.

Size: 1 to 1.5 cm (0.4 to 0.6")
Temperature: 15 to 30 °C (59 to 86 °F)
pH: 7.0 to 8.5
Tank size: 10 litres (2.6 US gal) and more
Distribution: South-East Asia

## *Neritina variegata*
### Tattoo Nerite

Also frequently found in the trade under the name of “Batik Snail”, since *Neritina variegata* and *Neritodryas* sp. (the “real” Batik Snail) are very hard to discern by the untrained eye. Both species are very variable regarding coloration and pattern, most frequently of a dark to yellowish brown with a black pattern. *Neritina variegata* and *Neritodryas* sp. show different behaviour, though. *Neritina variegata* remains under water most of the time, whereas *Neritodryas* sp. tends to climb out of the water, and most of the time also out of the aquarium, too. In the wild, *N. variegata* inhabits fast-flowing rivers and brooks. Its habitats are usually in pure freshwater, not in the tidal zone. However, like most Nerites, *N. variegata* does not reproduce in freshwater. The larvae drift into the sea with the current and spend several weeks there until the young snails wander back into the estuaries and then upstream, back into fresh water. In the aquarium, these snails are coveted algae-eaters. They also accept fish food flakes and food tablets, though.

Size: 2 to 3 cm (0.8 to 1.2")
Temperature: 20 to 30 °C (68 to 86 °F)
pH: 7.0 to 8.5
Tank size: 30 litres (7.9 US gal) and more
Distribution: Japan to Australia, Indonesia to the Solomon Islands

## *Neritodryas* sp.

### Batik Snail

Often imported as *Neritina* sp., which is not quite correct. They most definitely belong to the genus *Neritodryas*, which includes five known species. The snails that can be found under the name of Batik Snail in the trade belong to a hitherto undescribed species, though. They are very variable in colour, from yellowish to reddish brown, and have a batik pattern, black rings or interrupted lines. They are quite similar to *Neritodryas cornea*, whose natural habitat overlaps with that of the Batik Snail. Both live in brackish to fresh water and are often found on trees and bushes outside the water

around the estuaries of rivers and brooks. Basically it can be said that this is not an exclusively aquatic snail. It is thus not recommendable to keep them in aquaria, as they tend to leave the water as well as the tank and then starve to death somewhere behind the aquarium cabinet. When put in a tank they usually leave the water very soon, which makes them at best suitable for aquaterraria. In their habitat, some of these snails are found up to 1.5 kilometres respectively a mile away from the next waterbody.

Size: 2 to 3 cm (0.8 to 1.2")
Temperature: 20 to 30 °C (68 to 86 °F)
pH: 7.0 to 8.5
Tank size: 30 litres (7.9 US gal) and more
Distribution: South-East Asia

## *Neritina violacea*

### Red-Rimmed Nerite

Their shell is often covered in algae or lime deposits covering their natural colour, which may range from black and yellow or black and white to pink and white. The underside of the shell is mostly of a whitish yellow to orange or red. These snails are good algae-eaters, however, they need a large tank, especially when kept in groups. Their main food consists of algae and deposits on rocks. In the aquarium they can also be fed with powdered food. Please make sure it is not sucked away by the filter (it may be necessary to turn off the filter to let the

food sink). Red-Rimmed Nerites do not reproduce in fresh water, the larvae of *N. violacea* need brackish or marine water to develop. As in the habitat, these snails prefer rocks or other hard substrates in the aquarium. They often sit on the glass, grazing the surface for algae. As their natural distribution stretches as far as Japan, where water temperatures drop quite significantly in winter, they may be kept in unheated aquaria without problems. These snails also tend to leave the water, so keep your aquarium tightly covered.

Size: up to 2.5 cm (1")
Temperature: 15 to 32 °C (59 to 90 °F)
pH: 7.0 to 8.5
Tank size: 80 litres (21 US gal) and more
Distribution: South-East Asia, from India to New Caledonia and from Australia to Japan

## *Neritina turrita*

**Orange Track Nerite, Polka Dot Nerite, Zebra Nerite**

Frequently imported, however, their enormous variability makes identification extremely difficult for the untrained eye. They enter the trade mostly

under the name of *Neritina* sp. or they are simply called Nerite Snail. In the aquarium hobby, they have long been known as great algae eater, and they are recommended as first stock especially for large tanks. In nature, these snails inhabit the estuaries of brooks and even large rivers where you can find them on rocks and driftwood. Here, temperatures vary between 20 and 30 °C (68 to 86 °F). Especially on sunny days, 30 °C may even be surpassed. Aquaria should be tightly covered, as *Neritina turrita* has a certain tendency to leave the water. Outside the aquarium, they soon die of suffocation. Besides algae, they accept all kinds of foods from the trade, but they also like fresh food like dried red kuri squash (Hokkaido squash), brown oak leaves or red bell pepper very much. Their colour pattern may be serrated or even, with or without interruptions, yellow or black. Their basic colour is usually yellow, a dark reddish brown or even black. It is almost impossible to breed them in an aquarium. Even though these snails constantly lay their yellowish egg cocoons of about 1 millimetre in length on rocks, driftwood and even the aquarium glass their larvae do not survive in fresh water as they need brackish or even marine conditions to develop.

Size: 2 to 3 cm (0.8 to 1.2")
Temperature: 20 to 35 °C (86 to 95 °F)
pH: 7.0 to 8.5
Tank size: 50 litres (13 US gal) and more
Distribution: South-East Asia to Taiwan and French Polynesia, South and Central America, USA (from Brazil to Florida)

## MYSTERY SNAILS

Mystery or River Snails are widely distributed on all continents save Antarctica and South America. They belong to the Viviparidae, the live-bearing snails. Unlike so many other hermaphroditic snail species, River Snails are dioecious. All Mystery Snails are operculate, which means they have a trap-door-like lid (the so-called operculum) attached to the upper surface of their foot. This allows them to close off their shell entirely, which protects them from cold, drought or even predators.

## *Taia naticoides*
### Piano Snail

The Piano Snail usually has a cream-coloured, yellowish or greenish shell with brown stripes of varying colour intensity. However, there are also unicoloured specimens. Since the latter are deemed less attractive, they aren't brought into the trade very often. In their habitat, *Taia naticoides* eat detritus as well as biofilms and graze on rocks and wood. Most frequently, however, they are found in mucky and muddy substrates where they search for microorganisms. They do not eat higher plants. As substrate in the aquarium they prefer fine sand (as a replacement for mud) which they are known to dig into. *T. naticoides* are filter-feeders and should thus not be kept together with fan shrimp, since the latter are severe competitors for fine floating food. Feeding them with powdered food while the filter is turned off has proven beneficial. They also accept store-bought foods like flakes, powdered food and even small live food. *Taia naticoides* is not a robust snail and does not compete well with other snail species either, thus they should not be co-housed with just any other snail. Piano snails are dioecious live-bearers. Under good conditions, the females release offspring every two to three weeks. They can easily be sexed: The right tentacle of the males is thicker than the left one, and always a bit curled. These snails are quite susceptible to changing water parameters.

Size: 3 to 4 cm (1.2 to 1.6")
Temperature: 18 to 28 °C (64 to 82 °F)
pH: 7.0 to 8.5
Tank size: 50 litres (13 US gal) and more
Distribution: India to Vietnam

## *Viviparus acerosus*

### Danube River Snail

This species is on the IUCN Red List. In the trade, it is mainly sold as a snail for garden ponds. However, in nature it is found in different waterbodies. The Danube River Snail inhabits lakes, rivers, smaller ponds with muddy and sandy substrate or even rocky ground. There the snails feed on algal growth, detritus and mud containing large numbers of bacteria. In the aquarium they also accept flake food, spirulina, granulated food and food for fish larvae. As they are filter feeders they are not as easy to keep as other species. During the cold season they bury themselves in the substrate to hibernate. During this time, the snails do not eat, living off their reserves. After long, cold winters many of the weaker ones do not come out of hibernation. During warm winters, if water temperatures exceed a certain point, hibernation may be broken. Like all Viviparidae, the Danube River Snail is dioecious. The females release live, fully developed young snails. The common river snail, like all Viviparidae, is easy to sex, as the males' right tentacle carries their sexual organ, which makes it significantly thicker than the left one.

Size: 3.5 to 4 cm (1.4 to 1.8")
Temperature: 3 to 26 °C (37 to 79 °F)
pH: 6.8 to 8.5
Tank size: 50 litres (13 US gal) and more
Distribution: Central and Eastern Europe to the Ural Mountains, in temperate to moderately warm climate zone

*Viviparus* sp.

SNAILS

**RAMSHORN SNAILS** or *Planorbidae* are distributed throughout the world. There are those with shells shaped like the horn of a ram, for example *Planorbarius corneus*, however, there are also some species with bowl-shaped shells called limpets (formerly Ancylidae). The latter are not widely spread in the hobby, though. All in all, we have around 300 species in this family, 200 of which are “classic” Ramshorn Snails. Around 100 have a bowl-shaped shell.

## *Planorbarius corneus*

### Large Ramshorn Snail

The Large Ramshorn Snail has a dextrous shell with a reddish to dark brown colour. Its body is of a dark brown to black. It is one of the most attractively formed snails to be kept in an aquarium. Its preferred food are algae, especially green algae. The Ramshorn Snail does not eat plants, only dead plant parts, and as a scavenger it is almost indispensable in a planted tank. It is even said to eat blue-green algae. Please make sure to keep the population in check – when food is scarce they might start chewing on soft-leaved plants. Ramshorn Snails are highly suitable for propagation tanks where they eat surplus fish food that is left over from feeding young fish. Ramshorn snails are hermaphrodites, thus all *P. corneus* can (and will) mate with each other.

Size: 3 to 4 cm (1.2 to 1.8")
Temperature: 4 to 27 °C (39 to 80 °F)
pH: 6.0 to 8.5
Tank size: 20 litres (5 US gal) and more
Distribution: Europe to Siberia

## *Planorbella duryi*
### Seminole Ramshorn

Originating from Florida, USA, the Seminole Ramshorn is closely related to the European Ramshorn Snail but does not reach quite its size. These snails are especially sought after by aquarium keepers as they are available in many colours and shades of blue, white, pink, orange or yellow. In the lighter-coloured variants, the red-pigmented blood containing haemoglobin shines through the relatively unpigmented shell, which is, however, the only thing that discerns them from their inconspicuously brown relatives in Florida. They reproduce very well. Like all Planorbidae, the Seminole Ramshorn is hermaphroditic. Moreover, it is also autogamous, i. e. able to fertilise itself. Their reproduction rate depends very much on the amount of available food. If there is a lot of food in the tank (either put in by the owner or due to rotting plants), the Seminole Ramshorn reproduces at astonishing rates. The snails attach their transparent egg bundles to aquarium plants, other decoration materials or the aquarium glass. After around two weeks, 10 to 20 young snails hatch from each egg bundle. The Seminole Ramshorn is not a picky eater and will happily accept plant-based food as well as animal proteins, which makes this species very suitable in tanks as the snails do away with leftover fish food or dead shrimp and fish very reliably. Just a few minutes after one of their tank mates has deceased you can see the first snails munching on the dead animal. However, the Seminole Ramshorn does poorly in the presence of larger snails that are strong competitors for food.

Size: 1.5 to 2 cm (0.6 to 0.8")
Temperature: 15 to 30 °C (59 to 86 °F)
pH: 6.5 to 8.5
Tank size: 10 litres (2.6 US gal) and more
Distribution: in former times only in Florida/USA, today spread world-wide through aquaristics

SNAILS

# THIARIDS

The Thiaridae family is distributed in the tropics – and in aquaria throughout the world. Some of these operculate snails are trivially known as Trumpet Snails. Many Thiarid species are parthenogenetic livebearers with a characteristic and highly efficient brood care. The females have a special brood pouch where their eggs develop into young snails. Thus the offspring have a far greater chance of survival.

## *Melanoides tuberculata*

Malayan Trumpet Snail

*Melanoides tuberculata* is easily discerned from its relatives by its longish, sharply pointed shell that is closed off by an operculum, a trapdoor-like lid. In contrast to the other thiarids, the Malayan Trumpet Snail breathes exclusively with gills. Many aquarium keepers do not even know that they have these snails in their tanks at all, as they burrow in the ground and are only seldom seen by day. Most of the time they only leave the substrate at night. The Malayan Trumpet Snail is highly beneficial in aquaria because its burrowing activities help bring oxygen-rich water into the substrate and prevent it from solidifying. This is not only advantageous for the substrate itself and the bacteria that settle in it, but also for the plants in the tank, as nutrients are brought to the plant roots by the burrowing snails, too. The formation of rot spots in the ground is made practically impossible.

The Malayan Trumpet Snail mainly lives off dead plant parts, dead animals and whatever else it finds on the ground. It is not only a highly efficient scavenger and an inconspicuous aquarium inhabitant, but also a very good water quality indicator. If the water quality is in decline, the Malayan Trumpet Snail will leave the substrate and migrate to the water surface on the aquarium glass in great numbers. This is a sure sign that a water change is urgent. *Melanoides tuberculata* is quite variable in colour and form, it can be of a yellowish white over brown to

black, with a smooth or ribbed shell. In nature, they are found in stagnant and flowing waterbodies. *M. tuberculata* is a livebearing, dioecious snail, however, the females mostly reproduce by parthenogenesis. The young snails enter adulthood with a size of 1 centimetre and start reproducing then, too.

Size: 2 to 3 cm (0.8 to 1.2")
Temperature: 15 to 35 °C (59 to 95 °F)
pH: 5.5 to 8.5
Tank size: 10 litres (2.6 US gal) and more
Distribution: world-wide in tropic and subtropic zones. Has been introduced to temperate climates in the meantime, too.

## *Tarebia granifera*
Quilted Melania

This burrowing species is quite variable in shell colour, from cream-coloured over yellow, a light brown to a dark reddish brown, with purple stripes or dots. The shell surface also shows a variable knobbly pattern. The body of *Tarebia granifera* is of a dark grey with light-coloured spots or dots. In nature they are found in all kinds of waterbodies, from fast-flowing rivers to stagnant waters and even puddles that may attain a relatively high temperature in summer. The only thing they really do not seem to tolerate are low temperatures. In the aquarium, they are relatively uncomplicated. Their main food consists of dead plant parts, dead

animals and all kinds of fish food you can buy in a pet store. *Tarebia granifera* spend most of their lives burrowing in the ground, which is the reason why their keepers don't see a lot of them. They are dioecious, but also reproduce parthenogenetically. Females are much more common than males. The female Quilted Melania releases up to 30 fully developed young snails, which turn sexually mature after only three months.

Size: 2 to 2.5 cm (0.8 to 1")
Temperature: 15 to 30 °C (59 to 86 °F)
pH: 6.5 to 8.5
Tank size: 10 litres (2.6 US gal) and more
Distribution: India to the Philippines and Japan to French Polynesia

## *Thiara cancellata*
Hairy Trumpet Snail

As a rule, this crepuscular/nocturnal snail species disappears in the ground immediately after being put into a new tank. Once acclimated to the new pa-

rameters they can also be seen on the glass where they graze for algae and other microorganisms. Less than optimal living conditions and discomfort seem to cause them to intensify their burrowing activities. As substrate, these snails require sand or fine gravel that allows them to dig. If they feel comfortable they are visible during the day in the tank, too. The Hairy Trumpet Snail does not really go for most store-bought feeds except for those with a high Spirulina content. Tablet food made of pure Spirulina is preferred. Some keepers have noted that in their tanks, *Thiara cancellata* also eat brine shrimp, frozen fish foods and uncooked pasta. It is also highly probable that they are filter feeders and eat small particles from the water. *Thiara cancellata* is a dioecious snail. The females release larvae that drift to the sea with the current, where they develop. After a few weeks the young snails return to fresh water. Up until now, the Hairy Trumpet Snail has not been bred in captivity. Some specimens do not have the long hairy spikes on the shell, as they probably broke off when the snails were collected and/or transported inappropriately.

Size: up to 3 cm (1.2")
Temperature: 22 to 27 °C (72 to 81 °F)
pH: 7.0 to 8.5
Tank size: 60 litres (15 US gal) and more
Distribution: India to Indonesia and the Philippines

## *Thiara scabra*

### Spiky Trumpet Snail

*Thiara scabra* is a highly adaptable species, living in flowing waters as well as in lakes or swamps. In the aquarium, these snails are known to burrow a lot, however you will also see them on the surface quite frequently. They eat fish food as well as vegetables and detritus. To complement their diet, they graze for microorganisms and algae on aquatic plants, rocks or driftwood as well as in the substrate. *Thiara scabra* is a dioecious snail, however, the females also reproduce parthenogenetically. They release fully developed young snails every 10 to 14 days. One female may have 10 to 30 young snails. The spikes on the shells are a lot smaller in tank-bred specimens and often even nonexistent. The shells are of a yellowish brown with small brown or dark red spots. The foot is grey with a yellow spotted or dotted pattern.

Size: 2 to 3 cm (0.8 to 1.2")
Temperature: 20 to 28 °C (86 to 82 °F)
pH: 6.5 to 8.5
Tank size: 10 litres (2.6 US gal) and more
Distribution: Indonesia to Japan

## PACHYCHILIDS

The Pachychilidae family is also very widely spread in the tropics, and in earlier times, they were sometimes thought to be Thiaridae as they look quite similar. However, the families are not even closely related. All pachychilids live in fresh water, with the one exception of *Faunus ater*, which is also found in brackish waters. This family comprises oviparous (egg-laying) as well as ovoviviparous and viviparous (live-bearing) species. The females of the viviparous species have a special brood pouch where their eggs develop into young snails, which are released when fully developed. This gives their offspring a head start at survival.

## *Brotia pagodula*
### Pagoda Snail, Horned Armour Snail

The Pagoda Snail is dioecious and viviparous, but has a relatively small number of offspring. They reproduce only once a year, although their low productivity is compensated for by their relative longevity. They tend to do poorly in an aquarium, though, probably due to insufficient or altogether wrong food. *Brotia pagodula* is a filter feeder and thus needs very fine food in its tank. When keeping this beautiful snail in a dedicated tank make sure

to install a circulator pump in the tank besides the filter. Thus you can turn off the filter while feeding, and the circulator pump keeps the powdered food afloat. *Brotia pagodula* prefer cooler temperatures, clear water rich in oxygen and a good current. They are easy to identify, as no other snail species has pronounced spikes, almost horns, on the shell. The spikes on the shells of *B. binodosa*, *B. costula* and *B. spinata* are much smaller. However, the Pagoda Snail forms less pronounced spikes on the newly grown shell when kept in an aquarium. Its shell is of a reddish brown to dark brown, its body is basically grey with yellow spots.

Size: 5 to 6 cm (1.9 to 2.4")
Temperature: 18 to 25 °C
pH: 7.0 to 8.5
Tank size: 50 litres (13 US gal) and more
Distribution: Thailand and Burma

## *Brotia herculea*
### Giant Tower Cap Snail

This species is frequently offered in pet stores, however, not much is known about its behaviour. The shell of *Brotia herculea* is yellow to dark brown, its body grey with lighter-coloured spots. In their habitat, these snails are found on hard substrates

as well as on muddy ground and sand. The rivers and creeks they live in are clear and have a strong current. *Brotia herculea* frequently digs into muddy substrates and is thus often only half seen. They are a little less picky eaters than the other *Brotia* species currently found in the hobby, as they accept tablet food and blanched vegetables when kept in a tank. *Brotia herculea* is dioecious and has already been bred in captivity, although rarely. They only reproduce once a year, the female releases 60 to 100 fully developed young snails then. In the aquarium, the young snails require muck, algae and biofilms to feed on.

Size: 9 to 10 cm (3.5 to 3.9")
Temperature: 22 to 28 °C (72 to 82 °F)
pH: 7.0 to 8.5
Tank size: 60 litres (15 US gal) and more
Distribution: Myanmar and northwestern Thailand

## *Faunus ater*

### Black Faunus, Copper-Coloured Snail

Its shell is of a very dark brown to black with a coppery shine, which is the reason why one of the trivial names of *Faunus ater* is "Copper-Coloured Snail". Its body is usually grey with dark patterns, however, it may also be of a bright orange or even

yellowish. Black Faunus Snails are mainly found in the estuaries of larger and smaller rivers, where they graze on the muddy ground.

When the tide is low, they move towards the sea with the freshwater flow. We found them in a river only 50 metres from the shore. When the tide rises, the snails move in the opposite direction. In the aquarium they eat fresh vegetables, but also dry fish food or Spirulina powder. They often burrow in the substrate. Not much is known about the way they reproduce, however, their distribution implies that they have a Veliger stage that drifts into the sea where the larvae develop, to wander back to fresh water after metamorphosis. It is thus highly improbable that they will reproduce in an aquarium.

Size: 8 to 9 cm (3.1 to 3.5")
Temperature: 22 to 30 °C (72 to 86 °F)
pH: 7.0 to 8.5
Tank size: 80 litres (21 US gal) and more
Distribution: Western Indo-Pacific region

## ASSASSIN SNAILS

The representatives of the Buccinidae family are distributed throughout the world, from the tropics to the arctic, however, they almost exclusively live in marine waters. Some species are also found in brackish areas, and *Clea helena* lives in fresh water. Almost all species of this family are predators or feed on carrion.

## *Clea helena*

### Assassin Snail

Highly sought after in the aquarium hobby because it is carnivorous and feeds on other snails. The Assassin Snail even goes after snails far larger than itself. Assassin Snails have been observed to hunt and kill their prey in groups. This is probably not due to group behaviour, though, but to the fact that Assassin Snails are lured to search and attack their prey by the juices it secretes. Assassin Snails also eat the egg clutches of other snail species as well as soft-skinned invertebrates. Even planarians or other worms are eaten. Attacks on shrimp seem to be rare, although they do happen, and fish are not attacked at all. Often the snails' keeper will find them entirely buried in the substrate with only the siphon sticking out, which appears to be a localisation organ for potential prey. If there are not enough food snails in the Assassin Snail tank, *Clea helena* can also be fed with fish food high in protein. Assassin Snails are dioecious. The females lay egg cocoons with a single egg, from which a young snail hatches after three to four weeks, depending on the water temperature. In nature, these snails are found in all kinds of water bodies, from fast-flowing rivers to lakes with muddy ground, on hard as well as on soft substrates. Their shell is cream-coloured to yellow or brown and often has a light or dark brown spiralling stripe. The body and mantle are white or light grey with grey or black dots or spots.

Size: 1.5 to 2 cm (0.6 to 0.8")
Temperature: 23 to 30 °C (73 to 86 °F)
pH: 7.0 to 8.5
Tank size: 10 litres (2.6 US gal) and more
Distribution: South-East Asia

# POND SNAILS

The Pond Snails or Lymnaeidae are a family of pulmonate snails distributed in fresh water world wide. At the moment, this family is being revised and systematically regrouped. All in all it can be said that systematics are quite difficult here, as intraspecific variability is extremely great regarding shell forms and other morphological variations, which makes classification very difficult.

## *Lymnaea stagnalis*
### Great Pond Snail

*Lymnaea stagnalis* is a native European snail that is often introduced into the aquarium with live fish food. The largest whorl of the dextrous shell is bulbous, the rest elongate and sharply pointed. The entire shell is yellowish brown to dark brown in colour. This snail is quite often regarded as an unwelcome guest in an aquarium as it does not only eat algae and dead plant parts but, in its adult stage, it eats holes into practically all aquarium plants. One of its favourites is *Vallisneria*, whose leaves are not only perforated but quite often bitten off and eaten altogether. However, if you take out the adults that have become too large during regular aquarium maintenance you can keep the damage in check. The snails need to be constantly controlled though. *Lymnaea stagnalis* is a pulmonate snail, and hermaphroditic. In nature, sometimes its habitat dries out or freezes over, which does not really pose a problem. It also tolerates brackish water. The Great Pond Snail is thus even found in the northern part of the Baltic Sea. There it stays considerably smaller though, with a shell length of only 20 to 30 mm (0.8 to 1.").

Size: up to 6.5 cm (2.5")
Temperature: 4 to 27 °C (39 to 80 °F)
pH: 6.0 to 9.5
Tank size: 50 litres (13 US gal) and more
Distribution: Central Europe.

SNAILS

# BLADDER SNAILS

Bladder Snails, belonging to the family Physidae, are widely spread, with the two species *Physa fontinalis* and *Physa acuta* as their most prominent representatives. Sooner or later, almost every aquarium is colonised by them.

## *Physa* sp.
### Bladder snail

Most frequently of a dark colour and between 13 (*Physa fontinalis*) and 20 millimetres long (*Physa acuta*), i. e. 0.5 to 0.8 inches. These snails are – sooner or later – almost unavoidably introduced into every aquarium, usually together with plants. An interesting fact: *Physa* sp. is a pulmonate snail and has to go to the water surface to breathe. They are among the few species with a sinistral shell. Their preferred food consists of rotting leaves, but they also eat algae. Fish meat is only eaten when already decaying. Considering this we have to say that these snails are really very beneficial for the biological system of an aquarium as they could rightfully be called the tank health police since they scavenge on dead plants and fish, which they do away with in no time. Where there is too much food in the aquarium they may develop into a major pest. Restricted feeding or adding a predator to the tank (e. g. long-arm shrimp, crayfish or Assassin Snails) may help. It has also proven helpful to add natural snail bait like raw potato slices, which are not really eaten by fish, to the tank after the lights go out. The snails will gather on the potato slices and can thus be taken out of the tank in large numbers at a time. As a rule, bladder snails constantly lay egg clutches on plants, driftwood, rocks or the aquarium glass. Fully developed young snails hatch after two to three weeks, depending on the water temperature.

Size: up to 20 mm (0.8")
Temperature: 4 to 32 °C (39 to 89 °F)
pH: 7.5
Tank size: 5 litres ( 1.3 US gal) and more
Distribution: in almost every aquarium in the world

# ELEPHANT/RABBIT SNAILS

Snails of the genus *Tylomelania* became available in the trade only a few years ago. Their highly attractive appearance and their interesting behaviour have gained them a huge fan base within a very short time. *Tylomelania* snails are widely distributed on Sulawesi, Indonesia, however, most of these snails usually imported for keeping in aquaria originate from the Malili Lake System (Lakes Matano, Towuti, Mahalona, Lontoa and Masapi). Depending on their species, they live on hard or soft substrates. Several species share overlapping habitats in the littoral areas of the lakes down to a depth of 2 metres (6.5 ft). The snail densities in some places are extraordinarily high. All species originating from the lakes thus require high temperatures between 26 and 32 °C (79 to 89 °F), considering that the temperature of the surface water layer is between 28 and 32 °C (82 to 89 °F) all year round. Even in depths of 10 to 15 metres, or 32 to 49 ft. respectively, temperatures of around 27 °C (80 °F) are measured from January to December. Not only the temperatures are quite uncommon, but some other parameters, too. The Malili Lake System has a relatively high pH of 7.8 to 8.5, however, the water is rather soft (300 µs/cm). *Tylomelania* snails do not tolerate low temperatures well. Below 23 °C (73 °F), the snails withdraw into their shells, especially those collected in the wild. Among the genus *Tylomelania*, there are several really beautifully coloured species, many of them still undescribed. Unfortunately, not much information is available about them yet. Besides those species originating from the lakes, there are quite a few that come from the rivers on the island. Water temperatures there are generally lower than those of the lakes, only 21 to 23 °C (70 to 73 °F), which in turn leads to different keeping parameters, of course.

*Tylomelania* snails are dioecious, the young snails are released from the mother snail's brood pouch when fully developed. Only one young snail is released every few weeks. They are slow growers, an overpopulation of the aquarium is thus highly improbable.

## *Tylomelania* sp. "Yellow"
### Yellow Dwarf Rabbit Snail

In its natural habitat, this species inhabits shallow water zones and is mostly found on rocks and stones there. With a length of only up to 3 cm (1.2"), the Yellow Dwarf Rabbit Snail belongs to the smaller *Tylomelania* species. It is dioecious. The females release one fully developed young snail every six to eight weeks. Their shell is dark brown to black, their foot of a bright yellow to white, with grey lateral stripes. In the wild they feed on the bio-

film they graze off of rocks and driftwood, however, when kept in an aquarium they have to be fed in addition. A great food for them is powdered Spirulina evenly sprinkled on the water surface. When it soaks it starts sinking, and the snails can simply eat if from the ground. During feeding, the filter has to be turned off in order to allow the powder to sink. The Yellow Dwarf Rabbit Snail does not tolerate temperatures below 20 °C (68 °F), and they need a high or neutral pH.

Size: 2.4 to 3 cm (0.9 to 1.1")
Temperature: 24 to 29 °C (75 to 84 °F)
pH: 7.8 to 8.5
Tank size: 10 litres (2.6 US gal) and more
Distribution: Sulawesi, Lake Poso

## *Tylomelania patriarchalis*
### White Spot Rabbit Snail

These large snails are mainly found in muddy or sandy areas of Lake Matano, and hardly ever on rocks or driftwood. In the aquarium, though, they are also reported to search for food on hard substrates. Up until now they have not been observed to burrow. In captivity, they eat flake food, powdered food and tablet food, they do not go for live plants or dead animals. *Tylomelania patriarchalis* can be socialised very well with shrimp or unobtrusive fish. They are dioecious, and the female gives birth to a single, fully developed young snail every few weeks. At birth, the young snail is already 1.5 to 1.7 centimetres (0.6 to 0.7") long. The basic body colour of *Tylomelania patriarchalis* is black with a large number of small white or sometimes yellowish dots or spots. Its dark shell has a pronounced axial sculpturation. These snails do

not tolerate water below 23 °C (73 °F) and become gradually more inactive with falling temperatures. Under adverse conditions the females tend to have panic births.

Size: 10 to 12 cm (3.9 to 4.7")
Temperature: 27 to 30 °C (80 to 86 °F)
pH: 7.0 to 8.5
Tank size: 60 to 100 litres (15 to 26 US gal) and more
Distribution: Sulawesi, Lake Matano

## *Tylomelania drachenfelsi*
Orange Rock Snail

In Lake Poso on the island of Sulawesi, this snail usually lives at depths of 7 to 9 metres, or 23 to 30 ft. respectively, together with several other *Tylomelania* species. Here they can be found on rocks as well as

on muddy ground. They seem to prefer hard substrates, though. We found some young Orange Rock Snails on large rocks at a depth of only 1 to 2 metres, or 3 to 7 ft. respectively, in the lake, too.

Their dark brown to black shell is pronouncedly sculptured, spiralling as well as axially, its body is of a bright yellow to orange. In the aquarium, they eat flake food, powdered food, Spirulina algae and tablet food as well as fresh vegetables like red kuri squash (Hokkaido squash) or cucumbers. Like all *Tylomelania* snails, the Orange Rock Snail is dioecious and thus needs a partner of the opposite sex to reproduce. The eggs develop in the female's brood pouch, and the fully developed young snails are released, one at a time, every few weeks. Immediately after being born, the young snail goes searching for food. They grow quite slowly and reach a length of around 1.5 to 2 centimetres, or 0.6 to 0.8", six months after birth.

Size: 6 to 7 cm (2.4 to 2.8")
Temperature: 27 to 30 °C (81 to 86 °F)
pH: 7.5 to 8.5
Tank size: 60 litres (16 US gal) and more
Distribution: Lake Poso, Sulawesi

# MUSSELS

## MUSSELS

The mussels (class: bivalvia) are also a part of the mollusc family. During the last few years, more and more of them have made their way into the trade and consequently into our aquaria. However, only very few of the 1 200 species living in fresh water that have been described scientifically so far are found in the tanks in our homes. As mussels are usually completely covered by substrate in their habitat we show these invertebrates here in an unnatural position for photographic reasons.

Even in very acidic waters like here in Mc David/Florida, mussels can be found, which have especially adapted to this uncommon habitat so they are still able to build their calcareous shell despite the low pH.

Mussels are filter-feeders, and thus require a sufficient amount of fine nutritious floating particles like algae and microorganisms. With their filtering apparatus, they take their food from the water. For doing so, the mussel permanently pumps a stream of water through its body, which also supplies it with oxygen. Keeping mussels is not easy, as our aquaria tend to be much too clean and too well-filtered, and they simply starve to death if they are not especially fed, which is quite complicated. Mussels are very robust and tough animals, and thus they may even survive for several months under adverse conditions, but only because it takes them this long to starve to death – which is not an acceptable way of keeping any animal!

Those who are unwilling to make an effort to inform themselves more in-depth about the special requirements of these invertebrates and who are not ready to provide them with a dedicated tank tailored to their needs should really think twice before buying a mussel. Feeding them right is quite complicated if they do not sit in a dedicated mussel tank. It takes a lot of time to feed them fine-grained food through thin tubes in order to avoid fatalities. For doing so, it is best to use an infusion line from human medicine, most favourably equipped with a regulator. Mix the mussel food (finest plankton, algae, powdered food) with aquarium water in the infusion bottle, and position the end of the tube inside the aquarium with a thin stick so the water current brings the dissolved food to the mussels. All mussels are very susceptible to changes in the water, especially around their feeding siphon, thus the food needs to be about the same temperature as the rest of the aquarium water. In the beginning you may notice that the mussels at once close their shells if you come near them with the feeding device, however, they will desensitise after a time, given that you have enough stamina to keep trying, and feeding will become easier bit by bit. Many mussels are not sessile but move around in the tank, thus you need to keep your feeding device flexible. As you can see, it is possible to keep mussels in a tank, however, you cannot just buy them and leave them to fend for themselves in the aquarium.

In garden ponds, though, mussels make a lot of sense and are quite popular thanks to their filtering function. If possible try to get local species. Here in Europe, the Swan Mussel (*Anodonta cygnea*) and the Painter's Mussel (*Unio pictorum*) are the most suitable species. Unfortunately, in the trade you can often find the Chinese Pond Mussel (*Sinanodonta woodiana*), which must under no circumstances be introduced into nature. Among the species shown here, there is another one that is considered invasive. The Asian Clam (*Corbicula fluminalis*) poses a realistic threat to aquatic ecosystems especially in warmer waterbodies. It has already conquered many waters all over the world (Europe, USA and South America) where it is found in large numbers – with up to several thousand specimens per square metre.

The Zebra Mussel (*Dreissena polymorpha*) was introduced into Central European waters, too. It is especially unwelcome in swimming ponds due to its sharp edges. It occurs in very large numbers, and it does not need fish as intermediate host as it has, unlike most other freshwater mussel species, free-floating larval stages. It has a strong filtering function, however, it can clog waterlines, tubes and filters (also garden pond filters) as it reproduces in incredibly large numbers. Massive populations of tens of thousands of Zebra Mussels in several overlapping layers per square metre are not uncommon.

*Pseudodon vondembuschianus*

*Scabies crispata*

*Uniandra contradens*

*Corbicula fluminalis*

*Dreissena polymorpha*

*Hyriopsis bialatus*

*Parreysia burmana*

*Pilsbryoconcha exilis*

*Polymesoda sp.*

In the soft ground of the Blue Spring in Florida (USA) you can find a large number of mussels belonging to several different species.

# SOCIALISATION

Please make sure you research carefully before you socialise invertebrates with other inverts, fish or plants in order to avoid grave and possibly even critical errors. Without an exact knowledge of their requirements you will not be able to assess what these animals really need. If you choose aquarium inhabitants just like you choose the colour of your substrate or your backdrop, i. e. solely for aesthetic reasons, you will most probably run into severe problems and face utter disappointment sooner or later. Even organisms that live together in nature may cause trouble in the confined space of an aquarium.

Combining animals from different continents in one tank is another thing that is prone to go wrong. In addition, with the degree of relationship, the danger of transferring disease increases. Snails are far less prone to transmit germs to shrimp than e. g. crayfish, as shrimp and crayfish are much more closely related.

## Crayfish with other crayfish

Crayfish of different species should not be socialised at all. It may even be difficult to keep several specimens of the same species in one aquarium in the long run. If you socialise different species, even though they may share their habitat in nature, usually one species will dominate the other and suppress it. Besides the problems that may arise due to competition for food, hiding-places and space, another very grave danger is the transmission of germs between crayfish of different regions. There are crayfish species that carry a germ that won't do any harm to them because they have adapted to living with it very well, and yet crayfish of another species will suffer severe damage or even die when confronted with it. For this reason, crayfish from North America should not be socialised with those from any other region of the world. Crayfish Plague, a very virulent disease caused by a mould fungus North-American crayfish potentially carry infects crayfish from other proveniences without doing any severe damage to the original host, as a rule. For crayfish from regions other than North America this disease is absolutely deadly, though. Besides Crayfish Plague, there is a host of other bacterial and viral diseases much less well researched but still with potentially fatal consequences, especially for species that have never come into contact with these germs in their natural habitats.

## Crayfish with fish

Those who want to keep crayfish together with fish of course need to take the adult sizes of the future fish and crayfish as well as their feeding habits into account. Less robust species will be dominated by the stronger animals, and possibly end up as food. Most healthy fish are not actively predated upon by crayfish. There are few exceptions, and we have mentioned these in the descriptive texts of the respective species. Nonetheless, the basis for successful socialisation is a well-balanced diet and sufficient amounts of food. If fish or crayfish are very hungry or get a severely unbalanced diet they will eat just about anything they can get hold of. The smaller the crayfish (e. g. all species of the genus *Cambarellus*), the less problems you will run into. Of course you shouldn't keep these crays together with large predatory fish in order to prevent them from turning into fish food. Besides the danger of being eaten, some fish and crayfish compete for potential hiding places and caves. Thus you should never keep ground-dwelling fish together with crayfish as they will constantly squabble for the best places to hide. The larger the crayfish and the smaller the fish the more peaceful cohabitation will turn out to be.

## Shrimp with fish

Successful socialisation of shrimp with fish depends very much on the group the shrimp belong to. Fan shrimp for example can be kept in a tank with small peaceful fish without any problems, given that the fish leave the shrimp alone and do not try to nibble on their delicate fans or molest them in other ways. Most long-arm shrimp, however, are completely unsuitable for social tanks with fish as they are nifty and active hunters that can rid almost any tank of fish in a very short time. Only very few long-arm shrimp species with a rather small adult size can be kept together with fish. With dwarf shrimp, the situation is different, quite naturally. Shrimp are a highly welcome live food for many larger fish. Small fish will not eat adult shrimp though, and can be kept together with them. However, shrimp offspring may still turn into fish food due to their small size. Only very few highly productive shrimp species are able to establish a stable population in a tank with a relatively low number of fish (like e. g. *Neocaridina davidi*).

## Shrimp with other shrimp

It is also not recommendable to just socialise any shrimp species with another. Long-arm shrimp should never be kept together with other shrimp, for example. Besides some species with a small adult size, it is difficult enough to keep a pair of one species in the same tank for a longer period of time without them cannibalising each other. For them, dwarf shrimp are nothing but a highly welcome addition to their daily food. Fan shrimp, despite their larger body size, are also unsuitable for a long-arm shrimp tank as these peaceful shrimp will suffer damage or be eaten entirely during their next moult – at the latest.

Dwarf shrimp and fan shrimp can be socialised, however, freshly hatched dwarf shrimp offspring are potential live food for the latter, and survival rates are prone to decline. Different dwarf shrimp species kept in one tank will hybridise if they are closely related, with a more or less attractive outcome. Shrimp species that are known not to hybridise will still not do too well when kept together in the long run as sooner or later one species will dominate the other, and the suppressed species will slowly dwindle away and disappear entirely after some time.

## Crabs with other crabs

Crabs can rarely be socialised successfully with other crabs in the long run. Competition for food and space is fierce, and conflicts abundant, so one species will soon dominate the other, and the suppressed species will die off. Crabs should always be kept in a dedicated tank with no other inhabitants.

## Crayfish with shrimp

Keeping crayfish in the same tank as shrimp is possible, given that you choose compatible species. In many subtropic habitats, there are dense shrimp populations in the waters, and some of their members are eaten by the crayfish there. However, the shrimp compensate for this fact with a high reproduction rate. Socialisation may even work with less productive shrimp in an aquarium if you make sure you never keep small crayfish species like those of the genus *Cambarellus* with dwarf shrimp e. g. of the genus *Caridina*. Socialising larger crayfish with small shrimp is much more favourable. The presence of shrimp in a crayfish tank may even have very positive effects on the tank biology as dwarf shrimp are great for cleaning up after the crayfish have eaten. Large fan shrimp (of the genera *Atya* and *Atyopsis*) are often hurt or even killed by crayfish, though, especially after moulting. Long-arm shrimp are hardly suitable for social tanks at all, and most representatives of this group pose a critical danger even for crayfish larger than themselves. After moulting the crayfish will be attacked and severely hurt or even killed, if not before.

## Crabs with other invertebrates

Any attempt to keep crabs together with crayfish or shrimp will most probably not be successful. Even in nature, crab and crayfish habitats hardly ever overlap. Even small crabs will attack crayfish, and after the next moult the crabs will kill the crayfish for sure. The same applies to shrimp.

## Snails and mussels with shrimp

Molluscs (snails and mussels) and dwarf as well as fan shrimp can be kept together without any problems. Long-arm shrimp, in contrast, will regard snails as a highly welcome snack, and only highly productive species will last for a longer time when socialised with them.

## Snails and mussels with (or vs.) crayfish or crabs

Molluscs are a highly welcome additional food for many crayfish and crabs, too, with the exception of maybe the smallest species of the genus *Cambarellus*. However, even those should only be kept with robust snails with high reproduction rates. Large snails like Apple Snails can be severely damaged by overly intrusive crayfish as their breathing siphon often attracts their attention. However, some aquarium keepers deliberately use carnivorous decapods as regulatory factors in tanks with an excessively large snail population.

Some crayfish and long-arm shrimp have developed behavioural patterns that enable them to overwhelm and kill even large mussels. Especially *Procambarus clarkii* has been observed to wedge its large claw between the mussel's shells, trying to pry them open a bit, and using its smaller claws to attack and tear at the soft mussel tissue until the mussel goes limp and can then be killed.

## Shrimp or mussels with aquatic plants

Shrimp do not do any harm to healthy aquatic plants. Among the three groups, there are no species that are known to damage aquatic plants severely. The same applies to mussels, which may uproot a plant when digging into the ground but are otherwise completely harmless.

## Crayfish with aquatic plants

Among the crayfish, only the species of the genus *Cambarellus* and a few cave dwellers hardly ever damage plants, or do not do so at all. Most other crayfish tend to use plants as a dietary complement, and some of them are so diligent that no green remains in the tank at all. Even if you feed very generously they will keep nibbling on the plants. Some species are known for their burrowing activities, which damage plants from below by uprooting them. Large crayfish will just walk over plants, not around them, and thus simply snap them off. A good plant for crayfish tanks is hornwort (*Ceratophyllum demersum*) as it can be kept floating and need not be planted into the ground, or especially fast growers like waterweed (*Egeria densa, Elodea canadensis*). In some countries, these plants are considered an invasive species, and are thus illegal to import, sell, or even distribute, though.

## Crabs with aquatic plants

Among the crabs, there are also species that regard anything green in a tank as food. However, species of the genus *Geosesarma* can be kept in a planted aquaterrarium very well.

## Snails with aquatic plants

Among the snails, there are true "lawnmowers" that can develop a ravenous appetite especially for softer plants. Other species, however, only eat algae and other biofilm organisms, and thus fit into planted tanks especially well as they are a highly efficient clean-up crew.

# INDEX

## D



## E



## F



## G

# GLOSSARY

**adult:** sexually mature
**albinism:** without pigmentation
**antenna:** large tentacle of an arthropod
**antennule:** small tentacle of an arthropod
**aquatic:** living in water almost exclusively
**benthic:** adapted to life on the ground
**carapace:** hard shell covering the back, thorax and head of all decapods (ten-legged crustaceans)
**carnivorous:** meat-eating
**cephalon:** head
**cephalothorax:** head and thorax of the decapods
**chelae:** decapod claws
**cuticle:** skin
**detritus:** decaying parts of plants or animals
**dorsal:** referring to the upper side or back
**ecdysis:** moulting process
**emerged:** growing above the waterline
**endemic:** only found in a limited area
**exoskeleton:** hard shell of the crustaceans
**exuvia:** the shell a crustacean sheds when moulting
**gonopods:** the legs the males use for inseminating the females
**gonopores:** reproductive apertures of crustaceans
**herbivorous:** plant-eating
**juvenile:** young or not sexually mature
**marine:** living in the sea
**megalopa:** final larval stage in crustaceans
**morphology:** outward body form
**omnivorous:** eating food of vegetable and animal origin
**oviparous:** egg-laying
**ovoviviparous (or viviparous):** livebearing
**parthenogenesis:** "virgin birth"
**pereopods:** walking legs
**pleon:** the abdomen of crustaceans
**pleopods:** swimmerets, or swimming legs
**radula:** the rasping tongue of snails
**rostrum:** pointy appendage at the upper frontal end of the carapace
**salinity:** salt content of the water
**setae:** bristles or hairs
**siphon:** a tube snails use to breathe through
**stridulation:** creation of noise by rubbing body parts together
**submerged:** growing below the waterline
**sympatric:** two or more species sharing one habitat
**taxonomy:** the science of classifying and naming living organisms according to their body form
**telson:** central appendage of the 6th pleon segment
**uropods:** leaf-shaped appendages on the 6th pleon segment forming the tail fan together with the telson
**ventral:** on the underside
**zoea:** early larval stage of crustaceans

# Scape your tank!

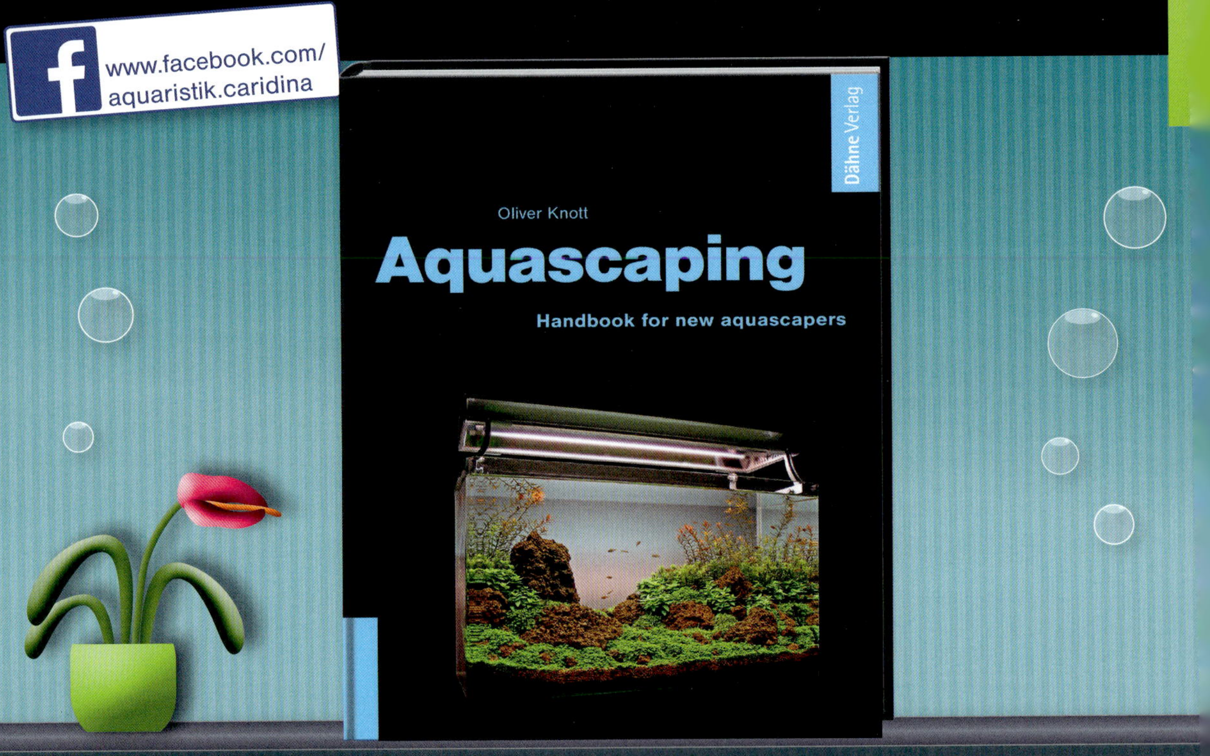


Oliver Knott

## Aquascaping

### Handbook for new aquascapers

96 pages, 200 photos, hardcover, ISBN 978-3-935175-93-7

Aquascaping is the modern way of designing artful aquarium layouts, of creating harmonious surroundings for aquatic plants and animals, and of setting up an aesthetically pleasing, attractive lifestyle accessory for your living room or office. The art of aquascaping is – quite unjustly – considered rather complicated and a lot of work. However, the famous, award-winning aquarium layout designer Oliver Knott shows in this book that even beginning aquascapers can create a great, easy-to-maintain layout with some basic knowledge and a bit of intuition.

**www.daehne.com/books**